Center for Intercultural Studies in Folklore and Ethnomusicology

The Folklore program at the University of Texas has come to be known, in the few years since its inception, for its distinctively performance-centered orientation to folkloristic teaching and research. It is thus with a special sense of appropriateness that we make available to the scholarly community of folklorists this English translation by James R. Dow of Mark Azadovskii's classic monograph, *Eine Sibirische Märchenerzählerin.* This seminal study, first published in 1926, represents a pioneering effort in the exploration of such performance-related problems as the community base of folklore performance, the organization of expressive repertoire, and the relationship between individual personality and performance form, all with a primary focus on the individual performer. These matters, which engaged few of Azadovskii's contemporaries, are very much before us now, though Azadovskii's monograph has long been out of print. The Center for Intercultural Studies in Folklore and Ethnomusicology is pleased to make this work available once again, in a form accessible to the growing numbers of English speaking folklorists, both students and professionals, devoted to the study of folklore as the creative expression of mankind.

RICHARD BAUMAN, *Director*
Center for Intercultural Studies
in Folklore and Ethnomusicology
University of Texas at Austin

A Siberian Tale Teller

MARK AZADOVSKII

Translated by
JAMES R. DOW
Iowa State University

Originally published as *Folklore Fellows Communications No: 68*
by the Suomalainen Tiedeakatemia

Academia Scientiarum Fennica

Helsinki 1926

Translated by Permission

Monograph Series No. 2

CENTER FOR INTERCULTURAL STUDIES IN FOLKLORE
AND ETHNOMUSICOLOGY

1974 * THE UNIVERSITY OF TEXAS * AUSTIN

Library of Congress Catalogue Card No. 74-26239
Standard Book Number 292-72006-8

FOREWORD

Since its inception in the early nineteenth century, the scholarly study of man as storyteller has been characterized by a curious ambivalence. Convinced that narrating was a "dying art" that survived only in remote areas and among socially backward or unsophisticated segments of the human population, the majority of pioneering researchers recorded, or urged others to "collect," tales known to nonurban and non-Western peoples in order to insure their preservation and to provide a data-base for historical studies of these seemingly soon-to-be-forgotten traditions. Because of their conceptions of, and attitudes toward, those from whom they elicited stories, most nineteenth-century fieldworkers, as well as the library scholars who scrutinized textual records made by others, showed little interest initially in the behavior of the narrators who served as their informants. Those who still knew and could tell stories, it was generally believed, were "tradition-bound" individuals who merely reproduced and transmitted to others ancient tales of which they were conceived to be both heirs and perpetuators; and the researcher's sole responsibility, it seemed, was simply to record these orally-communicated stories from any accessible and willing informants while it was still possible to do so. Consequently, little is known about the human beings who were the sources of most early folktales; and conceptions of the manner in which, circumstances under which, and reasons for which these individuals told stories were more often based upon investigators' presuppositions or assumptions about the ways of living and thinking of peoples whom they identified as "folk" or "primitive" than upon empirical evidence gathered by systematic observation or inquiry.

Despite these general tendencies, not all researchers were indifferent toward, or oblivious of, possible relationships between the experiences and idiosyncrasies of individual narrators and the tales they knew, told, and enjoyed. That such relationships must, in fact, exist seemed most obvious to fieldworkers who encountered individuals with extensive narrative repertoires or who had opportunities to observe and to interact with narrators who were dynamic or animated performers. Comparatively few in number and usually well known and respected locally as knowledgeable or talented artists, these master storytellers seemed to be more than merely bearers and transmitters of archaic tales they had learned from their progenitors or peers. Although the stories they told bore an unmistakable stamp of an old and persistent narrating tradition, they also seemed to be imbued with distinctive features indicative or reflective of the interests and preoccupations of their communicators. Consequently, a relatively small number of investigators was motivated to begin to inquire into the nature of the relationships they hypothesized must exist between the backgrounds and experiences of storytellers and their repertoires and narrating styles; and among the early fruits of such labors was a monograph written by the researcher Mark Azadovskii and published first in Russian and then in German in 1925 and 1926 respectively.

Translated here by James R. Dow for the first time in its entirety into English, *A Siberian Tale Teller* has been repeatedly and justly praised by folktale scholars as a pioneering and exemplary work of its kind. Committed to a research perspective initiated and developed, for the most part, by fellow Russian folklorists, Azadovskii was convinced that storytelling is a complex act, the nature of which cannot be fully comprehended or appreciated on the basis of the study of written records or "texts" of orally-transmitted tales alone. Folktales, he felt, are expressive manifestations of the personalities of their communicators and of the environments within which individual narrators learn and practice their art. Consequently Azadovskii took exception to the prevailing views that tellers of stories identifiable as *folktales* merely hand down or pass on to others, as faithfully or accurately as possible and with only minor or unconscious changes in content or style, narratives of unknown or obscure origins which have been preserved and perpetuated through time either exclusively or primarily by word of mouth. Instead, he conceived narrating to be a creative act, of which folktales are the artistic products or outputs; and the principal objectives of folktale scholarship, he asserts in his monograph, are "to determine those formative forces which govern the generation of a folktale" and to discover the underlying "artistic purpose" of storytelling. "The narrator," Azadovskii notes, "is faced, consciously or unconsciously, with the same assignment as the creative writer: the arrangement of his material, choosing and sifting the latter, and the formulation of his artistic intention"; and "the analysis of this artistic plan," he states, "is inseparably bound up with the study of the creative individuality of the narrator."

To demonstrate the nature and complexity of these interrelationships, Azadovskii focuses, in his short monograph, upon Natal'ia Osipovna Vinokurova, "a friendly little old woman of 50 or more," who told him tales "with obvious pleasure and with a noticeable pride in her narrative art." Following a brief description of the social milieu in which Ms. Vinokurova was born and lived (the Verkholensk district in the region of the Lena in Siberia), Azadovskii discusses in considerable detail the behavior of his narrator-informant, comparing and contrasting her repertoire and narrating techniques with those of other storytellers from the same district and demonstrating the ways in which and extent to which Ms. Vinokurova's tales and storytelling style reflect local customs and environmentally-determined concerns, on the one hand, and mirror personal values and biases, on the other. What emerges from Azadovskii's characterization is a portrait of a sensitive human being whose behavior in her role as narrator was dependent upon, and indicative of, perceptions, conceptions, and attitudes developed and shaped over time by environment and experience. Hence, to regard Ms. Vinokurova as a mere bearer and perpetuator of traditional tales, Azadovskii implies, is to ignore the distinctiveness of her artistry and to fail to recognize the fact that her narrating was both a manifestation and an index of her past interactions and personal preoccupations.

Although Azadovskii makes the behavior of a single narrator the focal point of *A Siberian Tale Teller,* two fundamental issues which he raises and to which he addresses himself are central to all studies of man as storyteller. The first of these concerns the apparent dichotomy between traditionality and creativity, on the one hand, and between communality and individuality, on the other. When an individual tells a story that he or she, his or her listener(s), or both conceive to be familiar, widely known, or frequently told, is the behavior of the person who narrates more accurately described as "traditional" or "creative," and is the story communicated better characterized as a "communal" or an "individual" phenomenon? Integrally interrelated with this is the question of the most defensible way to conceptualize the nature of narrating itself: is it a story-transmitting act during and by means of which some individual transmits to some other individual(s) a pre-existing phenomenon (story) that the narrator has either created or learned from someone else at some time in the past, or is it a generative process during and by means of which some individual formulates the story he or she communicates while interacting directly wtih some other human being(s)? A scholar's answers to these questions have a direct effect upon the kinds of research problems that he or she formulates and upon the kinds of investigative techniques and the conceptual framework or "model" that he or she employs. By posing these questions and concerning himself with the concepts and assumptions which underlie and are implicit in them, Azadovskii anticipated and articulated what have emerged as primary and controversial issues in twentieth-century folktale scholarship; and by advancing what is essentially a middle-of-the-road or compromising position in his answers to these questions, Azadovskii established a precedent that increasing numbers of investigators have either knowingly or unwittingly followed as inquiry into the nature and significance of narrating in human society has evolved and progressed.

ROBERT A. GEORGES
University of California
Los Angeles, California
July, 1974

Translator's Introduction

In the first lines of his Preface Mark Azadovskii indicates that there was a Russian original to his *Eine sibirische Märchenerzählerin.* The German text was published in Helsinki in 1926 by the Finnish Academy of Sciences and the Russian text in Irkutsk in 1925. Unfortunately,Azadovskii does not give the exact title of the Russian version and it took some insistent inter-library loan work and finally a letter to the Soviet Union to find out just which one of Azadovskii's many works this one actually was. When I did locate the original, what a pleasure it was. The text, *Skazki verkhnelenskogo kraia* (Tales of the Verkholensk Region), contains an introduction of about 40 pages, only 20 of which overlap with the German text of 70 pages. Thus the German is not just a translation, but is in fact a separate work, including a brief sketch of Russian folklore scholarship up to the time of the Verkholensk study by Azadovskii. The translation presented here is basically a translation of the German (FFC No. 68), but where the German and the Russian texts overlapped, the "double translation" was checked carefully. It is interesting to note that the Russian text, borrowed from the Indiana University Library, was apparently a copy sent by Azadovskii himself to Stith Thompson to be used in the 1928 translation and revision of the *Types of the Folktale* and the subsequent *Motif-Index of Folk-Literature* (1932-36).

There are a few places in the text where it was good to have both the German and the Russian, but there are other places where the Russian unfortunately did not supply what had been intriguingly hinted at in the German. Particularly I am thinking here about the reference on page 18 to those "expressions, proverbs, nicknames and contrivances of the Siberian village . . . which are hard to render in German." The Russian text is disappointingly lacking in details of this kind.

Two points should be noted about the present English text. All material appearing in brackets was put in by the translator, either to smooth out the translation, or to supply information not otherwise included. This includes, of course, the translations of the Russian titles in the Notes. Secondly, the system of transliteration used is the "modified" Library of Congress method; a key to this system of transliterating Cyrillic to Latin script has been included following the Preface. I have tried to be consistent in using this system, but often found it easy to overlook words which had already been transliterated in the German text. Little problems, such as *W.* Miller (in the German text) referring to *V*sevolod Miller, were very tricky and hard to pick up. I have nevertheless combed and polished the translation, and sincerely hope I have been as consistent and as accurate as I have tried to be.

Finally a word of gratitude to my colleagues who have helped me in the preparation of the translation. My thanks go to Robert Georges of UCLA who originally encouraged me to undertake the translation, to my colleagues in Russian at Iowa State University, Ann Vinograde and Tereze Michelsons, who helped me with the Russian when my knowledge of that language was lacking, to my former chairman Walter Morris for finding a quarter of released time so that I could do the work, and to Phyllis Soderstrum for typing the manuscript.

The work of Azadovskii, which every folklore student knows and most have struggled through in German, is now available in English. The man's interest in looking at the informant as a subject of study, and not just his products, in this case folktales, sounds very much like what is going on today in folklore scholarship.

JAMES R. DOW
Iowa State University
Ames, Iowa

Preface

The present work represents a revision of the introduction to a folktale collection made in the region of the upper Lena and published by me in Irkutsk in 1925. Soon after the appearance of this collection, the chairman of the Commission for Folktale Research (in the Russian Geographical Society), the academician S. Oldenburg, suggested that I publish for a larger circle of researchers and readers my findings on the exceptional poetic talent of the storyteller Vinokurova, whose folktales represent the contents of the aforementioned collection. It seemed all the more worthwhile since this collection of her tales had been published in such a distant place and because there had been only a limited number of copies. Several other specialists, to whom I turned for advice in this matter, shared this view. Among the opinions expressed, that of Professor W. Anderson of Dorpat was especially valuable. My hesitation disappeared completely when I became acquainted with a series of folktale investigations which had recently appeared in Western Europe, especially in Germany. Considering the interest which West European researchers are beginning to show so markedly in the personality of the singer or the narrator and his individual talent, it is my hope that observations I was able to make, when chance led me to the excellent Siberian narrator mentioned above, will not be without value.

At the same time I decided to offer a few general comments about material which has recently been collected by Russian researchers. This seemed all the more necessary, since the majority of these works was published in Russia at the time of the war and the revolution, and until now has not become the common property of West European scholarship.

To the lecturer in German at the University of Irkutsk, Herr Oscar Meissel, I would like to express herewith my thanks for translating my work into German.

Irkutsk, July 1, 1926 MARK AZADOVSKII

Key to Transliterations

В - V

Е - E Always e, regardless of whether initial or preceded by vowel, consonant, ъ or ь ; no distinction in transliteration between e and ё.

Ё - E

Ж - ZH

З - Z

И - I

Й - I*

Х - KH

Ц - TS*

Ч - CH

Ш - SH

Щ -SHCH

Ы - Y

Ь - ʹ

Э - Ė

Ю - IU*

Я - IA*

---

*The "modified" LC system omits diacritics; ĭ > i, t͡s > ts, i͡u > iu, i͡a > ia respectively.

A Siberian Tale Teller

I

IN CONTRAST TO West European research it is particularly characteristic of the Russian school of folklorists that they are interested in and pay attention to the personality of the singer or the narrator. This principle of individuality was first introduced into Russian [folklore] scholarship by the famous byliny collector of the Onega region, Professor A. Gil'ferding. To be sure, his predecessor P. Rybnikov had already pointed out varying presentational characteristics of the same material by various singers, but he had not yet succeeded in drawing all the resulting conclusions. The byliny in his collection are thus exclusively arranged by subject, even though the name of the singer, from whom this or that text came, is recorded.

The list of names became for Gil'ferding a beginning point insofar as it offered the possibility of setting up a series of controls. After ten years he undertook a trip, following in the path of Rybnikov: he visited the same places and recorded the same byliny singers. These written records presented researchers with most unique data for understanding the creative processes of the byliny. Older researchers, trained in the principles of the mythological school, saw in the byliny traditional and at the same time petrified and frozen products, an untouched heritage transmitted from one generation of singers to the next without change.

Gil'ferding's comparative investigations of the various written copies revealed, above all, an unusual possibility of change in the text. The same material not only assumed different form with the various singers, even the structure itself changed in time with the same singer. It turned out that formative involvement on the part of the singer was appearing with each new rendering of the byliny text.

Gil'ferding was able to see, that each bylina contains the "inheritance of his ancestors, the personal contribution of the singer, and the stamp of the region." At the same time the personal and basic element added by each singer plays the most important role. "Its importance," says Gil'ferding, "is much greater than one can imagine, especially when one hears the assurances of the singers themselves, that they are presenting the byliny just as they learned them from their elders."[1]

Gil'ferding presented convincing examples of how this reworking of the byliny text takes place. He has ascertained that every bylina is made up of two elements: that which is typical and to a considerable degree descriptive, or which consists of words set into the mouths of the heroes of the byliny; and secondly, by those transitional portions which bind the typical sections together and propel the course of the action. The first portions are learned by rote, the latter on the other hand are not; in the memory of the byliny singer only the skeleton is retained, and each time numerous variations are introduced: the same bylina can be recited differently each time by the same singer.

Even the typical portions, however, are not always the same: they have their own physiognomy with each singer, and present a reflection of his personality. "Each of these chooses from the mass of conventional epic scenes a larger or smaller supply, according to the strength of his memory, and then utilizes them in all of his byliny." Thus, for example, in the case of two singers, the *bogatyri* (i.e., the heroes of the byliny) are distinguished by outstanding piety, and each time, before every new undertaking, they pray to God for success. With other [byliny singers], however, the heroes are capable of wicked actions. Ilia, for example, normally God-fearing, shoots down golden church tower ornaments and crosses in order to pay the drinking bill of the drunkards in the tavern.

In one of the bylina about Alyosha and Dobrynia (Subject: the man at the wedding of his wife) the commands of Prince Vladimir to his servants, to keep out all those uninvited, are given in the greatest of detail, as are the conversations of Dobrynia with his servant. Thus, according to Gil'ferding, the "center of the tale has been placed in the antechamber." Such a rendering of the byliny was quite characteristic for its singer, a young man who had lived a long time in the city as a servant. In this way the impossibility of researching a subject, without preceding investigation of the circumstances which had brought about or influenced this or that change, came to light. Every circumstance mentioned by Gil'ferding suggested convincingly that the collected byliny should not be classified by subject when published, but rather by singer. Only by following such a system, the collector maintained, could one easily gain insight into the important question of the interplay between personal creative activity and inveterate tradition in the byliny.

Gil'ferding's theses were accepted by all later collectors, and the principle of classification and publication of the collection set up by him has become absolutely obligatory, even canonical, for all later collections. The same principle is used to classify the material in the collections of Grigor'ev (*Archangel Byliny*), Markov (*Byliny from the Region of the White Sea*), and of Onchukov (*Pechora Byliny*).[2]

This principle of arrangement began to spread then to other types of oral folk poetry. In the realm of folktale collecting this new trend in research proved to be effectual in the collection of Onchukov (N. Onchukov, *Northern Folktales,* St. Petersburg, 1908).

To be sure Onchukov had forerunners. In the eighties of the last century an outstanding collection appeared: *Folktales and Traditions from the Samara Region,* [St. Petersburg, 1884] assembled by D. Sadovnikov. The collector had carefully gathered biographical material about his informants. Unfortunately, however, he died before the appearance of his collection. The introduction which he planned and which was to contain this biographical data was thus never written down. Nothing is known about the fate of this important material. In the collection one finds only brief mention of the names of the informants along with the appropriate texts. The tales, however, were arranged by content. There are also a few biographical notes about the informants in various collections and articles, but these are only short notations without any special value and they have [thus] played a most unimportant role in the research of the Russian folktale.

For this reason the introduction of the principle of individuality into folktale collections must be traced back chronologically to Onchukov's collection. The same system of arrangement was carried out which Gil'ferding had set up for the epos. The locality principle is combined with the personal by Onchukov, i.e., the texts in his collection were classified according to the region from which they came, and within this framework according to individual narrators. Preceding the text collection of each folktale informant there is a brief characterization of him as a person and as a narrator. Incidentally, it must be stated that the characterizations often prove to be quite arbitrary and do not show sufficient depth.

The collection of Onchukov in turn became canonical for later collectors; almost all subsequent collections are arranged according to this plan and this principle. The collections which appeared during the World War in the publications of the Academy of Sciences and of the Russian Geographical Society have special importance for research on the Russian folktale and its bearers.

The war years and those immediately preceding represent on the whole a golden age of folktale collecting and research. In Moscow a number of young folklorists gathered around the person of the late professor and academician V[sevolod F.] Miller. The seminar which guided them stimulated a series of interesting theoretical articles and collections about folklore. In Petersburg the academicians A. Shakhmatov (now also dead) and S. Oldenburg assumed the leadership in research. Under the chairmanship of the latter a special commission for research on the folktale (the so-called "Folktale Commission") was established by the Russian Geographical Society. The main goal of this group was systematic collecting and publication of Russian folktales. Its first attempt was the publication of a special number of the journal *Zhivaia Starina* (Living Antiquity) to celebrate the one hundredth anniversary of the publication of the *Kinder- und Hausmärchen* of the Brothers Grimm. In this collection the previously mentioned Commission sketched out the inclusive plan and the broad scope of their assignments. Linked to the name of the Brothers Grimm, this work was distinguished by the variety of its material, which included the most diverse peoples of the extensive Russian territory. This work contained, in addition to the essays devoted to folktale research, the texts of Ukrainian, Great-Russian (among these Russian Siberian), Latvian, Lithuanian, Mongolian, Yakut and other folktales. The tales recorded by the famous Siberian ethnographer and collector A. Makarenko from a blind Siberian informant, called Chima, along with unusual characterizations of this man, contributed considerably to the embellishment of this collection. With the delicate gift of observation of an ethnographer and the sensitivity of an artist, Makarenko also sketches the region in which the blind informant tells his tales. Only [Dr. Ulrich] Jahn's unforgettable recounting can compete with this masterful description.

In the years 1914 and 1915 the Folktale Commission published successively the collections of Zelenin, the first of which was devoted to the *Great-Russian Folktales from the Province of Perm* [St. Petersburg, 1914], the second to those of the Viatka region. These two collections represent a most valuable document, both for what they have retained from the old folktale as well as for the changes to which they are subjected in the present. These collections also make a substantial contribution to the study of the importance and the nature of the informants.

The most important part of the collection of Permian folktales are the tales of the splendid narrator Lomtev. Twenty-seven texts came from him, of which some are so extensive that they require an entire quire. In the introduction, "Some Thoughts about the Folktales and the Story Tellers of Ekaterinburg," a detailed characterization of the folktale tradition of that region is given: the general tone [of the tales], the customs of the people, the conditions favorable to the preservation and diffusion of the tales, the groups of people who tend to become tale tellers, the context in which the collectors hear the tales which they record, the types of narrators, both male and female – these are the main questions which the author treats in his introduction. The folktale collection from the Viatka region is arranged according to the same plan and type system.

With these collections the Folktale Commission was concerned with isolating succinctly, in as far as this is possible, the individual elements of folk poetry, which were scarcely taken into consideration in former times and which were overshadowed by the idea of a folk-communal poetry – as is pointed out in one of the editorial conclusions.

Finally, in the year 1915, the monumental collection of B. and I. Sokolov *Folktales and Songs of the Belo-Ozero Region* [Moscow, 1915] appeared as a publication of the Academy of Sciences. The published texts are accompanied in this publication by an article of fundamental importance: "The Narrators and their Tales."

The Folktale Commission also published tales which were stored in the archive of the Russian Geographical Society. This publication appeared in two volumes under the editorship of A. Smirnov.[3]

Subsequent developments of life in Russia interrupted the work of the Folktale Commission and made impossible the publication of an entire series of collections which were ready for print. Not until recently has research and publication activity in this field begun again. The Folktale Commission has been active again since 1925. Somewhat earlier the East-Siberian Division of the Russian Geographical Society in Irkutsk began publishing folktale texts.

Thus, in Russian [folklore] scholarship, an immense body of texts and also of data about the life of the folktale and the role of the narrator has been assembled. From 1908, the time of the publication of Onchukov's collection, until today more than 1000 folktale texts have been published, and data for characterizing 135 narrators (100 men and 35 women) have been collected; moreover, there is a series of descriptions devoted to the customs and the milieu in which the tales are told. These descriptions treat chiefly the northern regions of European Russia (the provinces of Archangel, Vologda and Olonets). There is also similar material from the provinces of Pskov, Oreol and Tambov and from various parts of Siberia and the Urals.

The question of the dependence of the folktale on its narrator, as well as the question of the milieu, has also been touched upon by German scholars, even if only superficially. We find observations of this type in the excellent collections of Dr. Ulrich Jahn (*Volksmärchen aus Pommern und Rügen* [1891]) and of Wisser (*Plattdeutsche Märchen* [Jena, 1914]). The remarks of Ulrich Jahn, which agree in many points with the observations of Russian collectors, are especially valuable and are exceedingly important. However, both Dr. Ulrich Jahn and Prof. Wisser have stopped halfway, by not drawing all of the conclusions which follow. In their collections they follow the old system of arranging the material exclusively according to content. The *Märchen* told by one and the same person are scattered through the entire collection, partially as main texts, partially as variants, and often, thanks to the system of arrangement employed, they are left out completely from the collection.

These negative aspects of German *Märchen* collections have been pointed out by later researchers, for example W. Berendsohn, who writes the following in his treatment of this matter: "There are no collections which meet all scholarly research [requirements]."[4] The collections by Jahn and Wisser also have had no substantial influence on the further development of *Märchen* research in Germany. *Märchen* researchers leave aside completely the question of the influence the individual narrator has on the *Märchen,* or touch upon it only in passing. Even A. Aarne bypasses this question in his "Guide."[5] Only in the most recent works do German researchers give indication that they are beginning to grasp this problem in its entirety.

II

The material amassed by the Russian collectors and the observations made by them allow us to look more closely at the folktale and its development, to gain insight into requirements for the stability of tales in this region or that, and to ascertain the types of narrators.

The collectors all agree in stressing the importance of the narrator: "The personal taste of each narrator determines the choice of tales from the corpus [of folktales] which has been kept alive in the respective regions," says Onchukov; "the narrator retains in his memory, of all the things he has heard of this type, that which arouses his fantasy, stirs his heart or impresses itself deeply upon his soul."[6]

The collectors and researchers place more of their emphasis on the "biographical element." The tale contains deep traces of personal experience and of the events in the narrator's life. His experience, his ability to observe, his occupation with this or that handiwork, as well as his own personal characteristics, all of this has an enormous influence on the retelling of a tale. Especially the collections of Zelenin have an abundance of such examples. No doubt one of the most outstanding narrators is Lomtev (from the province of Perm), from whom folktales have been recorded for the last twenty years. He likes particularly to tell tales about merchants, which he presents according to his own recollections. Merchants overshadow everything else; even Ilia Muromets[7] appears for a short time in one of his tales as a merchant's servant in a shop. The narrator Selesnev from the Viatka region is an artisan. "And in his tales one can even observe the attempt to influence, in this way or that, the relationship of the employer to him as a worker. Thus, for example, the diet of the felt boot makers at the home of the forest demon is not bad for village circumstances: fresh wheat bread, and sweet creamy milk."

The personal attributes of the narrator are also reflected in the moral tones of the tales. The narrator Petrushchev for example, as the collectors call him, a man with a codex of morality concepts, with a predilection for fixed bearing and piety, prefers to stay with moral themes. "Examples of triumphant virtue touch him so deeply that he is genuinely moved by them." The choice of his tales attests to the direction of his thoughts and of his feeling. Thus the tales of Ivan the Fool, Nikolaus the Miracle-worker, and others. "Even in tales with a rather coarse content, as for example, 'The Merchant from Vologda' or 'The Little Groom,' the emphasis lies in establishing virtue." At the same time, in accordance with the gentle character of the narrator and his good nature, there is never any mention of punishment in his tales. Such a meeting of moods (as the Brothers Sokolov called it) appears most clearly in the tales told by women. In regard to this point, there is an abundance of material in the collection of Onchukov.

We find similar observations by Jahn. Dr. Ulrich Jahn shows with striking examples how the same *Märchen* can take on a different tone and a different coloration with various narrators.[8] But his observations are concerned chiefly with the content aspect and with the changes to which the role and the character of the main figures are subjected. The observations made by the Russian collectors, especially by the Brothers Sokolov, make it possible to grasp even the formal, stylistic changes which come about through the individuality of the narrator. An especially obvious example is the narrator Ganin. The latter is not only an outstanding folktale connoisseur, he is also one of the few preservers of the epic heritage which has been preserved in the region of Belo-Ozero. Knowledge of byliny has also influenced his folktale repertoire. The majority of his tales approach in content the motifs and themes which we encounter in Russian epic poetry. He told, for example, the tale of the marriage of Prince Vladimir, of King Solomon, and others. In his repertoire a broad stream of fabulous religious legends [*Märchenlegenden*][9] surfaces, which are related to sacred epic songs in character. That is the basic content of his tales. Byliny style elements and epic poetry technique are also characteristic of his tales. In them one finds commonplaces (*loci communes*) in abundance and turns of phrase which correspond exactly to expressions in the byliny recited by the same narrator. "From this," say the collectors, "it is evident what role the personality of the narrator can play in the process of assimilating folk poetry, even in the case of a subject which has come by chance into a certain region, or which was taken from a book."

A religiously inclined narrator fills his tales with allusions to God, with quotations from prayers and the Bible, with religious slogans, etc. The tale of a cynical narrator abounds in oaths, invectives and episodes of indecent character. There are narrators with a passionate love for tricks and jokes, and a serious magical tale takes on the character of a funny tale [*Schwank*] in the hands of these narrators. There are poetic tale tellers, and others with a rich fantasy who happily dwell upon the images from that magical world; there are those who dwell on their descriptions of magical objects with exquisite details, and there are narrators of the realistic type, who demonstrate a definite preference for descriptions of folk customs and of daily life.

The introduction of various local motifs into the tales is, in the final analysis, closely bound up with the personality of every narrator. It is what A. Aarne commonly calls "acclimatization." His formula: "Every folk imprints, so to speak, the tale with its own stamp," must be considerably expanded: with each folk [group] the folktale receives its individual, local character, which is added by the individual narrator.

We find extensive material concerning this question with all the collectors. In the tales from the Pechora region we find in almost every text forests, lakes, rivers and the ocean. The heroes stumble about in the impenetrable forests, hunt in the swamps, and travel across the rivers, lakes and oceans. Even in short, humorous tales mention is made everywhere of rivers and oceans.

At the same time all the common occupations of the region have been portrayed in the Pechorian tales, chiefly fishing and fur trapping; the forest and occupations associated with it prevail in the tales recorded in the province of Olonets.

The Ural, under which is to be understood not just the mountain range, appears as the chief locus of activity in the Permian tales. "Ural" means here every wilderness, every unsettled region. Below we will have opportunity to introduce analogous phenomena in the tales recorded in Siberia.

This is the range of the most important observations, which have been made by Russian collectors of the present. I will not go into detail here: there is rich material available which has as its object the context in which the tale is told, the folk customs, the conditions of preservation and diffusion of the tale, the reciprocal relationship between narrators and audience, the attitude of the narrators themselves to the tales, especially their belief in the miraculous, etc. These questions, for the moment, lie outside the realm of our specific interest. For me the primary concern was to sketch out the range of the most important questions which had been put forth by Russian researchers, as well as the most important observations and conclusions made by them.

III

At the same time a very obvious gap in Russian research must be pointed out. By placing emphasis on ethnographic elements (under which the personal is also to be understood), purely artistic questions were completely overshadowed. By the same token, questions concerning the formal structure of the folktale were not examined closely enough, nor were they treated with the necessary seriousness.

This phenomenon can best be explained by the general condition of Russian literary scholarship: Russian folkloristics has always been little concerned with stylistic elements. It followed in this respect the common tradition of Russian literary history, whose interests were limited for a long time to the realm of cultural-historical or psychological-biographical questions. On the other hand, pure literary history research was limited chiefly to questions of borrowings and influence. Incidentally, Russian literary scholarship was not the only one which had to undergo such a period. We run across a similar phenomenon in German literary history, even though it is not quite as glaring there.[10]

Folktale research also moved in the general direction of literary scholarship. One cannot help but agree with the author of one of the newest works, when he says of the onesidedness of scholarly Russian folktale research: "All works about the folktale are actually limited to comparative investigations of particular folktale themes [motifs]."[11] In contrast, questions of the formal structure of tales, stylistic studies of this type of story, and investigations of the folktale as a self-sufficient, artistic organism were unknown to Russian scholarship. That which was called "stylistic study" in several investigations, was in fact limited to a description of isolated technical peculiarities. Thus, completely new material which appeared in new accounts was treated in the same traditional method.

Methodological advances, which Russian literary history experienced together with West European scholarship,[12] also had a positive effect on folklore research, especially in the area of the epic and of lyric poetry.

A basis in structure and style studies is to be found in that which was accomplished earlier in the areas of theme history and that of cultural-historical research. Questions of poetics move into the foreground.

In my opinion, this problem appears with special emphasis in folktale research. The new accounts and the accompanying biographical notes about the narrators have literally opened up for us a new world. But this material has not yet been sufficiently expanded. Study of the personality of the narrator proceeds completely apart from [studies concerned with] more inclusive investigations of the world of the folktale. Basically the narrator has been of interest to us up to now only insofar as the knowledge of his uniqueness, of his personal characteristics, of the traits in his world view, and of biographical moments contribute to the discovery of this or that aspect of the folktale (tale motif). The narrator and his art, as a specific problem, has not yet awakened the interest of the researcher. The study of the style of every narrator is made only in conjunction with the changes which are made by him in the traditional subject matter.

Style study, however, is not just a means of ascertaining this or that stratification in the tale. In the style the entirely original nature of the folktale is revealed, [it is] a completely artistic organism. One must stop looking at form as something purely external and mechanical. It is not simply a summation of means of expressions, but is rather its organic system, a complete chain of phenomena arranged according to definite laws. Style research in the realm of the folktale must not be satisfied simply with an enumeration of individual artistic elements, but must accept as its goal, the ascertaining of meaning for each and every form element in the artistic whole of the tale. The observations which have been made by the collectors of the present about the life of the folktale and the role and importance of the personal elements in these [tales], present us [quite] directly with this problem.

The relationship between the tale and its narrator is determined primarily through its form. One must not imagine, as some researchers do - among these is Dr. Ulrich Jahn - that the personal element appears only in the changing of the milieu, in replacing one character with another, or by introducing local characteristics and facts from personal experience, etc. It is also not sufficient to point out the introduction of this or that characteristic trait borrowed from other spheres of folklife or from literature by this or that narrator. The importance does not lie in these details and in those elements which insinuate themselves from the external world (from that of the narrator). The goal of research is to determine those formative forces which govern the generation of a folktale. The narrator is faced, consciously or unconsciously, with the same assignment as the creative writer: the arrangement of his material, choosing and sifting the latter, and the formulation of his artistic intention.

Thus we must, to a certain extent, look at each tale as a completely artistic organism.

The folktale, like every other product of art, has as its basis a definite artistic purpose. This [artistic purpose] serves as a structuring basis for a series of pictures, the examination of which reveals all the individual elements and small stylistic details.

The discovery of this artistic purpose, the analysis of this artistic plan appears thus to be one of the most imporant tasks. This task is inseparably bound up with the study of the creative individuality of the narrator. Thus the problem of the importance of the personality of the narrator in the creative process of the folktale, suggested by the Russian school, needs to be studied in more depth. Narrowly conceived biographical studies must make room for the study of the artistic physiognomy of the narrator, his repertoire and his style.

It is very characteristic that several groups of German scholars have approached this problem. They have drawn the same conclusions, even though they have approached it differently. In contrast to Russian research a considerable number of works has been published by German scholars on technique and on the external form of the folktale. Especially valuable were the contributions made by the group studying with Prof. Kaufmann (from Kiel). The further development of these studies, however, makes necessary a substantial expansion of the range of goals set forth. This became most obvious in studies on folksongs, where the research of Murko, Lach and others made clear the insufficiency of formal-stylistic observations and the necessity of transferring ones attention in research to the conditions essential to the life of the folksong and its performance. It is precisely the same with folktale research. West European, especially German, scholarship is concerned with the themes [problems] set forth by the Russian school. Here as well as there the problem of the narrator is in the foreground. In my opinion the new goals are formulated with the grestest precision in the work of W. Berendsohn: *Basic Forms of Folk Narrative Art in the Kinder- und Hausmärchen of the Brothers Grimm*. "We need," – writes the researcher – "collections which present us with the living art of the folk, completely uncorrupted, with all its preferences and its mistakes. It is incumbent upon us to learn from the environment and the atmosphere in which narration takes place, from the relationship of the narrators to their audiences, from the contents of narrative art. Above all, however, we would like to get to know, through detailed descriptions, the narrators as well as their entire repertoire of stories, in order to interpret the importance of the personality in the restructuring of the *Märchen*; for we are no doubt dealing with highly gifted people with rich phantasies and strong memories, with artistic types who are the equals of outstanding creative writers in the world of literature."[13]

Thus, the study of the creative personality of the narrator is looked upon as one of the next and most important goals; the study of individual repertoires should accompany the study of the folktale as an entity. The present sketch will attempt to acquaint the reader with the creations of one of the most outstanding representatives of this world of creators and bearers of the folktale, the Siberian tale teller Natal'ia Osipovna Vinokurova.

IV

I became acquainted with Natal'ia Osipovna Vinokurova during my stay in the Verkholensk district where I had been sent, by the Folktale Commission mentioned above, for the purpose of collecting folktales. At that time she was a friendly little old woman of 50 or more, still very vigorous and energetic in her work. She agreed quite obligingly and without any affectations to my request that she tell me some tales. She told them with obvious pleasure and with a noticeable pride in her narrative art.

With naive pleasure she accepted the praise I bestowed on her and told me tales for days, without paying any attention to the dissatisfaction of her nearest relatives and to the ridicule of her neighbors. Only housework forced her frequently to interrupt her tales.

Vinokurova is completely illiterate. She has spent her entire life in the village where she was born, having lived in the city as a servant girl only for a short time in her early youth – a situation which, as we shall later see, had an indirect influence on her texts.

She picked up her tales from all over, as she said, wherever she had the opportunity to hear something. Several tales she heard in the city, but most she had learned from "knowledgeable people" among the inhabitants of her region and from travelers.

It must be stated that the region in which she lives is probably very rich in storytellers. I say this only as an assumption because Siberia as a whole, and the region of the Lena in particular, have been researched very little in regard to folk traditions.

Planned research did not start here until the end of the 19th century. But even that which has been accomplished thus far suggests to me that a rich epic tradition has existed in Siberia in the not too distant past. Rich folk lyric traditions and ancient customs have been preserved down to the present day. The folktale penetrated into Siberia together with these types of folklore, [all of] which were brought by the first colonists to Siberia and which have been carefully preserved by them.

The folktale has been well preserved in the Verkholensk region. In the relatively short time of my stay I was able to record from informants about 100 folktales and a series of tales and memorates of the recent past. I also collected reports about storytellers from other villages which I did not visit. I do not mean to say that the folktale on the Lena leads as full and as intensive a life as, for example, in the northern part of European Russia. However, it is still quite clear that the folktale on the Lena has not only not been forgotten and has not disappeared, but it lives on with the people and draws the attention of the listeners as well as the narrators. And, there are among its bearers outstanding masters to be found.

A whole series of objective conditions contribute to the preservation of folktales among the people in this region. In the first place must be mentioned that various jobs are carried out communally by the people. Agriculture is exceedingly difficult in this area. Thus the main occupations appear to be hunting, transporting of goods, travelers and mail, lumbering and long-boat building. The latter is one of the chief crafts of the old landed populace; there are many villages which earn their livelihood almost exclusively with this profession. Long-boat building is pursued especially on the Kulenga, a tributary of the Lena.

This handicraft is pursued chiefly by cooperating societies; the same being true for transporting of goods, passengers and the mail. And in the postal stations, in the hostels, and in the forest during communal lumbering for long-boat building, one often has the opportunity to hear tales being told during the rest periods. The narrators are well liked and are gladly taken in by the working groups. Almost all reports about tale tellers begin with the words: "We were in the artel,"[14] or: "We were out hunting. . . ." I might add, that hunting also seems to be a profession which is often carried out by cooperative groups, since the hunter is often compelled to go into distant forest preserves. The mill, as everywhere, is quite naturally a favorite place for listening to and telling tales. It is not just chance that one of my most interesting informants was a miller (in the village of Anga).

The presence of vagabonds (*brodiagi*) and the penal colonists also represents a very important factor. If one does not understand this situation clearly, it is difficult to understand the structure and the character of the folktale of these climes.

These two phenomena of Siberian life, or more correctly, of the Siberian past - the penal colonists and the vagabonds - are closely connected with the Siberian Deportation Institute, which has been in existence for many years. Penal colonists are those criminals who were sent, after serving their time, to some Siberian village where they had to be accepted into the village community.

In most cases this element could have little salutary effect on the development of the region. Even though many of them settled into their new lives, started housekeeping and after much effort and with much hard work acquired citizenship in this land so new to them, the majority was nevertheless incapable of work. As the author of a famous work on the penal colonists and *brodiagi,* N. Iadrintsev, says: "The majority of them had been deported because they have proven to be incapable of work and of civilization in their homeland, in European Russia." Furthermore, life in Siberia, especially in the eastern portions, is exceptionally difficult, especially in regard to the climate and the land. "It is considerably easier," continues the author, "to look for an easier way to make money, deception, thievery, begging, etc." It was here that the phenomenon developed which we call *brodiashestvo* (vagrancy).

A *brodiaga* is a deportee, a settler who has left the place to which he was assigned and has started out for home or is simply wandering around the country like a vagabond. Vagrancy was one of the most obvious results of deportation in old Siberia. According to the reports of Iadrintsev, which were made in the 60's of the last century, up to 4000 *brodiagi* wintered in the vicinity of Irkutsk in various judicial districts. Approximately 50 *brodiagi* on the average passed through the villages along the main highway in the summer, and about 20 during the winter. The total number of *brodiagi* in Siberia in the 1860's was 20 - 30,000. The region in which the tales of Vinokurova were recorded was also filled with *brodiagi.*

It is of course understandable that such a large group of people had to develop its own habits and customs, its own notions of law, and its own literature. All of this affected in one way or another the existence of the Siberian peasant and especially his poetic legacy. We present a particularly obvious example from the work of Iadrintsev mentioned above. "Occasionally, especially in the winter, when one comes into the vicinity of a village at night, one can see smoke climbing from every bathhouse. There the *brodiagi* have found asylum for the night. The peasants gladly give these wanderers shelter and even treat them with respect as long as they do no damage. The peasants like to listen to the tales and anecdotes of the *brodiagi.*"

These meetings and associations are not always accidental and temporary. Often there are cases of *brodiagi* staying for longer or shorter periods as workers for the peasants. Many of the narrators whom I had the opportunity to hear, named as their teachers *brodiagi* who had lived in the village as workers and from whom they had heard their tales. The same thing has been recorded by the Siberian collector Vinogradov. The famous ethnographer S. Maksimov tells of the former illustrious *brodiaga* and robber Gorkin: "Later he reformed, served his time, became an exemplary landlord and was known in all of Transbaikal as a postal courier (*Iamschchik*) and a skilled singer of songs. Finally, however, he could stand it no longer, gave everything to his son and went off begging. Visiting villages and huts, he loved to delight the children with tales and funny stories."

Some observers of Siberian folklife, for example Astyrev the famous publisher who lived for a short period in Siberia as a political outcast, were of the opinion that all Siberian folktales come from the penal colonists. This view is of course extreme, but it is certain that an entire series of stratifications in the Siberian folktale must be traced back to the influence of the penal colonists and the vagrant element.

But these stratifications are of a special type. Usually when a folktale penetrates into a new region, it [the region] first has an influence on the person of the hero, who assumes a new social quality. Ivan the czar's son in the old tale must make way for Ivan the peasant's son, the soldier or the farm-hand. In the tales of the deportees this is not the case. One must not forget that the position of the penal colonists, in the eyes of the old landed

Siberian populace, was different from that of a soldier or a farm-hand. While the latter, for example, was surrounded by a special aura of human emotions, a certain bravura, the colonist was looked upon as a pariah, often as something despicable, a buffoon, etc. Between these two groups of the populace existed a constant antagonism, a mutual distrust and an attitude of enmity. It is not accidental that a series of expressions, proverbs, nicknames and contrivances of the Siberian village have been preserved which have the penal colonists at the butt and which are hard to render in German.

It is quite evident that the penal colonist could not replace the former hero of the magical fairy tale, or even of the first novella-like tales. This would not have been accepted by the listeners and could not have been preserved in any way for a period of time. But in humorous tales in which it is a question of sly rascality or derision, the hero is now and then "a servant from Rasseia" (as Russia is pronounced in Siberian dialect). Thus, for example, in a tale I recorded in the village Kistenevo, which concerns a trick played on a sly Jew and his daughter, the hero appears as a fellow from "Rasseia" who lives as a servant with the Jew. He seduces the daughter by appearing before her at night wrapped in a sheet, with lights on his head, and claiming he is God. Occasionally, however, the nickname "Rasseiskii" (literally - a man from Russia) is given to a stupid or dumb figure in a tale. Moreover, one can see in this last phenomenon a reflection of a later situation, when a large settlement of colonists from European Russia took place toward the end of the last century in the far reaches of Siberia. The reception of the settlers on the part of the old landed populace was less friendly, and the latter have until today maintained a scornful and derisive attitude toward the new arrivals.

The deportation element is also seen insofar as specific motifs and details of [this] life are introduced into the tales; the appearance of vagrancy motifs is an example of this. Very often the hero, in various difficult life situations (for example when he has to escape from unjust punishment), vanishes into the forest and leads a "vagabond life" or simply "wanders about." One must also count the frequent reference to prison and the knowledge of prison society as a part of this.

But it is not in this that we find the principal meaning of the deportation element. It does not lie so much in the content as in the form. It is expressed in an especially clear manner in the composition of the tale.

Story telling for the penal colonist is not simply a pastime, not just a means of diversion, but is rather an important means of earning a living, in a certain sense a handicraft. Thus the necessity of giving the tale a maximum of exciting, interesting and clever moments. Thus the super abundance of obscene elements, which suddenly appear quite unexpectedly in the tale, thus the multitude of complicated entanglements of the subjects, thus the introduction of many episodes which often take on a meaning all their own.

I will recount one of the recollections I recorded about narrators [which I found among] the multitude of penal colonists. As far as my memory serves me, I will give the account just the way I heard it. "You are sitting in the evening at the gate, a colonist walks up. This and that, and then he begins to tell a story. Where does he get all that! And he talks on and on about it. It is getting late, but he keeps on talking. At times you don't want to let him go. Sometimes he takes two days to tell a story." Thus he is practicing the sly politics of Scheherazade. In order to earn his bed and his evening meal, he has to tell his tale in such a way that he captivates the none too receptive Siberian peasant. But the most important thing is that he must understand how to draw out the telling of the tale so that it is time to eat supper. Such tales and such facts [as these] are not isolated cases; they are also mentioned by the other collectors.

This peculiar construction of the tale: this multitude of plots and episodes, the occurrence of obscene and coarse elements, characterize completely the tales of the informant Anan'ev from the village Anga. Especially characteristic [for him] is the introduction of obscenities into the tale. Usually such a tale is constructed as something complete, a whole tale. In the center is to be found an appropriate subject, and in accordance with this, details are gathered together and worked out. Examples of such tales can be found in great quantity in the collections of Onchukov and of the Brothers Sokolov. There are even specialists just for this genre. [I] count one of the story tellers I heard in the village of Kistenevo, Alexander Malarov, as one of these [specialists].

The tales of Anan'ev, [as mentioned above], have a different character. With him the indecent element forms a portion of almost every tale. At those places where every other informant would have been satisfied with a brief hint or a cursory remark thrown in, Anan'ev picks up this hint and develops a broad, protracted and detailed episode.

One could speak of individual preferences of the narrator, but from my notes it becomes obvious that such tales, or more accurately stated, the manner in which they are told, is not unusual on the Lena. Most likely we are dealing here with a definite tradition, with a definite manner, with a special school.

Thus, we observe in the basic elements the influence of the vagrants and the penal colonists in two directions: in the form [structure] and to a limited degree, in the content.

The creations of Vinokurova take place outside of this colonist-formal tradition. In content, however, many customary factors are reflected, among them the case of vagrancy, e.g., in the tale The Son Captured by a Kiss (The Milk of Wild Animals, number 315 in Aarne's *Verzeichnis der Märchentypen*)[15]: the defamed wife of the czar sits imprisoned in a stone tower. A sympathetic servant helps her to freedom. "If your husband comes back, you won't live anyway. Wander about in the world with your son," says the servant. With his help she saves herself and goes out into the distant world. The king returns home. His minister, thanks to whose slandering the queen was locked in the tower, reports to him: "Your wife is roaming about in the world; I was not able to keep her in prison." The servant girl in the tale likewise saves herself by going out into the distant world. The Cursed Garden (Aarne no. 301). Locked in a stone tower because she had not protected the king's daughter, whose safety had been entrusted to her, she also saves herself by fleeing and wandering about in the forests, mountains and swamps.

It is possible, that the vagrancy motif was not brought in so much by the penal colonists themselves, as it was brought to life by the very fact of its existence [i.e., the penal colonies]. In summarizing this element, however, one must include the prison motif and the detailed knowledge of prison life in the Lena folktales in general and in the Vinokurova tales in particular. In the Tale of the Faithful Wife (Aarne, no. 882) the heroine sets out in search of her husband who was thrown into prison as the result of a lost wager, and thus she tries to get permission to see him again, "With various requests she turns to the police and to other officials." She goes to the prison warden. The latter absolutely refuses her permission to see her husband, reveals to her, however, that she might see him when the prisoners are being led into the bathhouse, just before their execution: "They led him into the bathhouse. She goes there too. The watchmen do not let her through, but she bribes them and she obtains permission for a meeting 'for fifteen minutes.'" It must be pointed out that we have here a faithfully and artfully drawn picture of life.

Along with this there are other times when [local] customs are presented without anything else. Closely bound up with the vagrancy motif, this wandering about, is a motif which we might designate as the sheltering of the wanderer. For villages which lie directly along the highways or which have in some way direct contact with them, this motif is very characteristic and acquaints us with one of the typical aspects of local life. In the tales mentioned above, The Cursed Garden, we meet the servant girl with three heroic sons, who has already become rich and who keeps a hostelry. "And one time travelers came and knocked at their door. The sons ask their mother whether they [the travelers] should be let in. She answers: 'Oh, dear children, let the travelers in. I myself wandered about for a long time and thus have pity with all who do." In the tale about the wise woman the son comes to his mother, with his newly acquired [purchased] wife dressed in poor clothes. The mother who believed her son to be dead, sits and weeps bitterly. "Just put us up for the night," asks the son who was unrecognized —"I don't let any wanderers in, I have enough grief of my own." The motif of taking in of a wanderer is also mentioned in the tale of the rich Marko. "Once a wandering beggar came to the house. He asked for lodging for the night. The woman refused and said: "Don't take him in, I am pregnant. Something can happen that will be unpleasant.' But the husband answers: 'Now, now, silly woman, what does it matter. We have a special room.' And they let the old man in."

On the whole, Siberian life flows like a broad stream through the creations of Vinokurova. Here it is seen in some detail of the customs, there it appears in some characteristic of the dialog, and the next time it turns up in one of the broad descriptions of typical Siberian manners and morals. Vasia, the son of the eagle prince, disguised as a girl, charms *koshchei*[16] with his play. The latter sends a servant to invite the would-be musician to an evening festival. "The servant asks the boy who is pretending to be a girl, but she (Vasia) answers: 'I will not be able to entertain your master with my performance – I come from the simple people, I am a simple *cheldonka* (a Siberian peasant).' " In the tale The Goose with the Golden Egg, the boys return to their father after one has become czar and the other a statesman. Not recognized by the father, the czar asks him: "Are you from Rasseia, grandfather, or from here?"

In the tales of Vinokurova we see almost all the occupations of the Verkholensk region: ferryboating, commercial transport, and hunting, among other things. In the tale about the rich Marko and Vas'ily a certain [ferryboat] pilot, the father of Vas'ily is mentioned: "He was a [ferryboat] pilot with an unusual fate. All the rich people knew him. . . ." We find a typical picture of the Lena in the Tale of the Wise Woman, [we see] the recruiting of workers for the gold mines and for the ferryboats. "Here is a notebook, Vania, recruit a hundred workers, write down the first and last name of each one, send them to me and advance each one of them 100 rubles." Very often the hunting profession is mentioned. It is very characteristic that the hunter's cabin has completely replaced the "little hut on chicken legs," which appears so often in the Great-Russian folktales.[17]

I will not linger with similar examples. I will only include one more scene which is typically Siberian in an especially obvious way. A servant girl of the czar, who has fled from the anger of her master, gives birth in the forest to three warriors. "And they say: 'What shall we do here? We have nothing with which to begin [our life].' And so the oldest thinks: 'Listen brothers, let's pile up tree trunks and roots on the road so that people can neither walk nor drive past.' Which they did. They blockaded the highway. And when wagon trains loaded with goods came along this road, they couldn't get past. 'Hire yourselves out to clean up the road, dear brothers.' Then they hired out. They are very strong and take on the entire job of cleaning up the road, for a hundred rubles. And inside of three hours they have cleaned the entire road."

As one can see from these examples, the tales of Vinokurova are in close association with local traditions and are completely permeated by local elements. However, the importance of the tales of Vinokurova does not lie just in this local coloration. In my opinion this informant can be counted among the best representatives of Russian folktale poesy.

V

Among the many informants, both men and women, whom I had the opportunity to hear in the Verkholensk district, one can distinguish three major types of narrators. Representative of the first type is the miller Anan'ev mentioned above, in whose tales the penal colony element is seen so clearly and completely, [especially] in the formal structure. As a representative of the second type we might consider the seventy-five year old Medvedev. This is a most important tale teller for whom the exact reproduction of the tale and every detail of the same is of importance. He narrated without haste, thinking about every detail at the same time and carefully observing the effect brought about [by telling of the tale]. He did not have a large repertoire, only four tales, but his texts were so inclusive and detailed that they were equal, in number of pages, to ten of Vinokurova's.

For such narrators this careful adherence to the traditional form of the tale, to that which one calls the folktale canon, is quite characteristic. He presents very strictly all the introductory and closing formulas, the obligatory epic law of threefold repetition, and so forth. The subject matter with such narrators is usually very unified and undivided, and only seldom does one encounter such combinations and concatenations of subjects, as one finds, for example, in Anan'ev's tales. In his narrative style the informant Medvedev is very much like Aksamentov, also an outstanding connoisseur and master. But his tales come mostly from the realm of the military and show clear evidence of life in the casern.

Vinokurova occupies a special place. Although her tales and those of Anan'ev come essentially from the same soil and are permeated to a similar degree with local Siberian color, it is difficult to find two more differently disposed folktale informants than these two. In contrast to the tales of Anan'ev, which are to a certain extent without order, which contain many subjects and are rich in episodes and details strewn about, there are the tales of Vinokurova: weak in details, unified and rather narrowly conceived.

That limitless disorder of the plot which we see so often with Anan'ev is not only lacking with her, she usually seems to have little interest in the plot itself. One hears often from her a comment, a rectification of the type: "I forgot to tell that." Sometimes she would not make this correction until the next day, or the day after that. Also, she does not place any great value in details of a fantastic nature.

This playing down of the plot is especially noticeable in her attitude toward the epilog. With her it is in most cases arbitrarily shortened and has little motivation. It becomes quite noticeable that she wants to end somehow.

She is different, as we shall see further, from the informants of the second type (the Medvedev and Aksamentov type), in her lack of interest in the folktale canon.

Let's take as an example a piece from the Tale of the Eagle Prince (Aarne no. 313B & 303). In rough outline the subject is the following: Ivan the merchant's son goes on a hunting trip, sees a wounded eagle and is about to kill him, spares him, however, at the eagle's request, and takes him home. Feeding the eagle proves to be so expensive that the merchant chases his son from his house. The eagle takes him on his back and flies to a city where his [the eagle's] oldest sister lives. There the eagle, with Ivan's help, takes on human form, and [he, Ivan] is sent by him [the eagle] to his sister. "Go up to the window," he says to him, "and ask for alms, not in the name of Christ, however, but rather for the sake of the eagle prince." The outcome of the tale is presented by the narrator in the following way: "He goes up to the window and asks for alms, not in Christ's name, however, but for the sake of the eagle prince. A servant girl was standing at the window ironing clothes. Now, and she ran, as fast as her legs would carry her to her mistress. 'How does one ask today for alms!' The mistress understood her meaning and walked up to the window herself; he tells her everything, and [then] asks for the keys. She listens to everything and then says: 'However long it has been since I have seen my brother, even if it were that long again, I would not give you the keys.' He returns to the eagle and repeats the words of the sister. 'Well, here it didn't work, let's go to another city to my second sister.' To make a long story short, there they also received a negative answer. They went to the third city to the youngest sister. Again Ivan the merchant's son went and asked in the same way for the alms. This [sister] was pleased immensely. 'Where is he then, the eagle prince?' –Give me the keys, then I will lead him to you.' She handed the keys over to him. And then they came with the eagle, the tale goes on, and then they celebrated. And then the eagle prince married Ivan, the merchant's son, to his sister. 'And I,' he said, 'will go and seek my fortune.' To Ivan he entrusted all twelve storehouses, in them was much gold and silver."

Let's stay with this example for a moment. Above all, a good folktale informant who follows tradition would have repeated exactly and in detail three times in the same way the request for alms, also the two negative answers of the sisters, But epic repetitions of this type, which would have been absolutely necessary for example for Anan'ev or Aksamentov, are completely lacking in the style of Vinokurova. She does not like to extend the length of the tale "without need." Each repetition of this type appears in her tales only as an unusual exception. The statement "in short," with which she joins and generalizes about a whole series of episodes in the example above, represents one of her favorite transitional and connective formulas. "In short, they took the bride on the very next day and sent her off home," and so forth.

Let's return to the example above. It is characteristic in another way. In one little section a whole series of occurrences is reported. The reunion with the sister, the marriage of Ivan the merchant's son, (incidentally sometimes he is called Ivan the merchant's son, sometimes Ivan the czar's son), the entrusting of twelve storehouses, and so forth – all this is presented in short, compressed, simple recounting of the facts, while the description of the customs and of the surroundings receive considerable attention. The time of the unexpected appearance of the brother, the clashing of sisterly love with avarice are both of more concern to the informant than all those purely folktale-like details of the marriage of the hero.

She is not interested in that which is magical and fantastic, but is interested exclusively in the local customs and in the psychological side of the folktale. At the very center of her interest there is a definite psychological episode, the unexpected appearance of a long lost brother, and she tries as carefully as possible to present *how that all could happen.* And in this fashion a colorful genre picture is drawn: a stranger walks up to the window, asks for alms, a young servant girl is ironing clothes at the window, she is frightened by the unusual "new way" of asking for alms, runs as fast as she can to her mistress, etc.

In the conclusion of the tale, this artistic method is presented even more clearly. The eagle prince sets out looking for adventure, after he has married his benefactor to his youngest sister. He arrives in the capital city of a strange land. "In this city lived and ruled the immortal *koshchei.* And he had kidnapped a merchant's daughter, whom he then kept. For some time the eagle prince lived in this city, and he visited the '*koshcheika*' (the wife of the *koshchei*), when the *koshchei* was not in the city. And this *koshcheika* became pregnant by him. And once the *koshchei* surprised the eagle in his palace and chopped his head off. And she was pregnant by him. And when the *koshchei* was away, she gave birth during his absence and did not know what to do with the child. For sure the *koshchei* would kill it. And she happened on the idea of placing the child in an oaken barrel. She wrote on the barrel that there was an unbaptized child within, and then she cast it into the ocean." What an abundance of folktale-like happenings – and, at the same time, there is a complete lack of artistic connections, a simple accounting, a bare registering of the facts! But in other respects the tale returns to reality, into the surroundings of everyday life. Ivan, the merchant's son, dreams that new ships have arrived in his port. Unsettled by the dream, he goes to the shore to see if everything is as it should be, he finds the barrel, opens it, discovers the boy in it and takes him into his house. We hear further how the boy grows up, plays with his comrades, learns in school that he is superior to all the other boys in physical strength and in his intellectual abilities, he learns that he is a foundling, etc. All of this is told with great care, with an exact description of detail. In short, as soon as the tale returns to the realm of everyday life, it again takes on formative and artistic details. Thus, Vinokurova always tries to bring the folktale-like surroundings close to actual reality.

Most characteristic of her tales is the complete absence of the usual, traditional, introductory, fantasy type formulas: "in some czar's kingdom," "in some distant land" . . . "over behind thirty lands, in the thirtieth kingdom of the czar," etc. Her beginning is always realistic and leads right away into the circle of the central figures and of their actions: "Once there lived an old man with his wife" . . . "A czar had three sons" . . . "Once there lived a king with his wife, and he had to go on a trip" . . . "A czar had no children" . . . "An old woman had a son, and he went to the fair" . . ."A soldier finished serving in the military" . . . "A merchant's son reached the age when one marries" . . .

E. Rohde set up the following formula for the development of interest in the folktale: "As is well known the folk does not at all like to be restricted to the narrow circle of its wearisome, poor life, not even during the relaxing flights of fantasy, during which they want to rest up from difficult work: where the true folktale does not rise above everyday limitations through *ironical treatment* of these limitations, then it prefers to go off into the blue and into a fantastic and grand existence. It is quite at home with kings and princesses, but to be sure these are folktale kings who move about so easily and comfortably, and talk as if they did not wear a mighty gold crown on their heads day and night."[18]

Even though this formula is truthful, nevertheless it must be supplemented. At a certain stage the tale begins to seem oppressed by this removal from reality, by this fantasy "world of kings and princesses." It tries to return to the "narrow confines of workaday life," to give meaning to these fantasy surroundings which have become strange and have gone astray, and through this the tale acquires new interest.

At the same time this weaving of fantasy with real life is expressed by different informants in different ways. With some informants real life is completely under the domination of fantasy. As one Russian researcher so accurately noted, "the fantasy details are so closely woven into those of everyday life that it is impossible to separate the one from the other." "Transferring alternately in his tale from the description of unusual happenings to a detailed presentation of a modest daily life, the informant is trying unconsciously to explain to himself the events in a fantasy life which have become incomprehensible, and at the same time he surrounds the everyday details with a certain mysterious veil.[19]

In the case of other informants, fantasy elements evaporate in real life. More accurately stated, the subject itself remains fantastic, evokes wonder, but the material for its development, for the solution of the one or the other problem of composition, are drawn exclusively from the observations of everyday life. At the same time material from realistic folktales and even from folk anecdotes is taken over into the poetics of the magical tale.

To this group of story tellers belongs Vinokurova. It is characteristic, for example, that *koshchei* the immortal has lost all his fantasy trappings in her presentation. He appears much more as a dignified lord who goes about his daily business. Not by chance does he plead in the epilog with the son of the eagle prince who has acquired the egg in which his death is enclosed: "Give me this egg, take over my position and I will go away from here." In a similar way in the tale The Czar's Daughter Freed by a Soldier (Aarne no. 300) the traditional monster of the tale is represented by a man whose height reaches "to the ceiling" – the only fantasy characteristic in this picture. Even the forest demon (in the Tale of the Unfaithful Sister), who has become the lover of a young girl, does not possess a single fabulous attribute. The boar's tooth and the designation "forest demon" – that is all that has remained from the original folktale pattern. The number of such examples can be expanded considerably.

Likewise one can observe gaps and a lack of clarity in the presentation of instances of fantasy in the tales of Vinokurova. Such gaps are quite obvious, for example, in the tale The Untold Dream. The informant forgot to mention the punishment which awaits the suitor, who can not solve the riddle, the final taming of the recalcitrant woman, etc. Her attention, as we have already said, is drawn to instances of real life and psychological impetuses. By and large the Vinokurova tales can be described as psychological, and that gives them their peculiar, special place, not just in the Siberian, but in the whole Russian repertoire.

This psychological coloration determines both the composition and individual stylistic characteristics in the tales of Vinokurova. Löwis of Menar has pointed out the special place of dialog in the Russian folktale. In the opinion of this researcher the essential difference in the artistic structure of the Russian and the West European folktale lies in the dialog.[20]

And, in fact: the individual mastery of the narrator comes out most clearly in the structure and the character of the dialog. The dialog of Vinokurova, in this respect, follows the general tradition, but in it a whole series of important peculiarities appear which give her a special place in the lists of informants.

As a rule a folktale informant almost never gives "stage directions." That is easy enough to understand. He is not an author, but a narrator. He presents the subject or an episode not only in words, he enlivens it through his own gestures, his own pantomine. Vinokurova, however, introduces gesture into her tale. A special place is left for the description of the gesture, and this then takes on a separate meaning. The dialog is supported by descriptions and movements and by pantomine, and often – and in this we see a high level of artistic mastery – through this allusion to a single movement, to a single gesture some occasionally complicated, inner experience of the hero is characterized.

Ivan the merchant's son finds the eagle and promises to feed him: daily one wether. The merchant's son takes the eagle and delivers him to his father. He tells all. *The father is quiet:* "that is expensive!" he says. In the same tale the son of the eagle prince sets out seeking the death of the *koshchei.* During his wanderings he comes into a region "where one can neither buy nor rent anything, and he becomes very hungry." "He has only a moldy crust of bread which he is about to dampen in the ocean and then eat. But scarcely has he walked down to the shore and dipped the crust of bread in when a fish comes swimming up and tears the bread out of his hand. He accuses the fish of robbing a traveler of his last bit of sustenance, *shrugs his shoulders* and goes on his way."

A still finer psychological detail, that also accompanies the dialog, we find in the same tale when the mother of the hero is interviewed by *koshchei.* "The son charged his mother with asking *koshchei* where his death was hidden. The next morning she invites the *koshchei* to drink with her. *Koshchei is very pleased about this.* "She has never loved him – and now she invites him to drink tea with her." And then, during tea, she begins to question him. At first roundabout, and then she asks him straight out: "How is it my dear, as long as we have lived with you we have never talked with you so earnestly. What sort of pleasure is it for you to give these parties, to tire yourself out, and see, now you are all tired out. Tell me, my dear, where is your death?" *Koshchei had to laugh:* "What do you need my death for?" . . .

The gesture is not always mentioned, but it becomes apparent in the tension and the vehemence of the dialog between the czar and his still unrecognized son. The czar asks him: "Can one not interrupt music and song? I have important things to talk over with you. Where are you from?" – "I don't know myself," he answers, "but I have heard my mother was a wash woman, my father – a cook." – "Cooks and wash women never bear such children. Tell the truth!" – "Well then, can a czar and a czarina beget dogs?" the son hastily interrupts. (The Wonderful Son)

Finally, in a variant of the famous Tale of the Unfaithful Sister, a colorful picture of complicated inner experiences of the hero is created in sharp outline by the use of gestures and movements. I will not include the quote here, since I must spend time with it later in connection with several other instances.[21]

This inclination of the informant toward psychological portrayal in the composition of her tales comes out with still greater strength. I have already pointed out the lack of superficial piling up of subjects and episodes in her tales. I have also mentioned her disdaining of fantasy requisites and epic details. This lack of repetitions and obligatory details, together with the lack of superficial motifs and episodes, gives her tales something monolith-like, a certain unity and completeness.

She puts two or three episodes in the foreground, stays with them and spins them out in detail. At the same time these episodes, according to the role which they play in the traditional material, are not at all those which are central [to the tale]. In the development as well as in the execution of these episodes, the interest of the narrator is not directed toward the external motivation, not toward a strict succession of facts, but rather *at the inner impetus of the action.* And even here she develops unusual characteristics of an observing psychologist.

These characteristics come out in considerable relief in the tale The Magician and his Pupil (Aarne no. 325):

The characteristics of this subject are well known. J. Bolte and G. Polívka present them as follows: "The father delivers his son to the magician as an apprentice, however, after a year recognizes him in his animal (or bird) form. The boy secretly learns magic and escapes. He has his father sell him as a dog, ox, horse,[22] finally [he is sold] to the magician, to whom the father, against orders, gives the rein. – He succeeds, however, in casting off these reins and wins the magician over by means of a battle of metamorphosis; usually, after he [the boy] has fled as a bird to a princess and is hidden by her in the form of a ring, the magician appears as the doctor of the ailing king [and] asks for the ring; when the king's daughter throws the ring down, a great quantity of millet seeds [appears], which the rooster then begins to peck at; but suddenly the youth becomes a fox and bites the rooster's head off."[23]

This is the usual scheme. The episode with the sale is presented according to Russian tradition in almost all variants in the same way. Two times the old man sells his son and two times the latter returns. The third time the father sells his son with the rein (or with the cage), and the son falls, as we said above, under the power of his earlier master, the magician. But the causes and the conditions of this last sale are individualized by the narrators. The usual explanation is this, that the old man is seduced by the high purchase price. In the case of Afanas'ev, for example, (no. 140, Variant d): "Sell it with the rein, and I will pay you more," says the buyer. The old man let himself be seduced, and he sold it." The same with Chubinskii (Vol. II, no. 102): "The buyer offered the seller as much money as the latter can lay hold of." With Rudchenko: "The gypsy adds five rubles for the rein. The old man considers this: 'The rein costs all together thirty copecks, and he is paying five full rubles,' [but he] can not resist the temptation and leads the horse away with the rein."

In the variant by Afanas'ev the old man simply forgets to take the rein off, likewise with Iavorskii (no. 36). In the Permian collection by Zelenin (no. 59) these two instances are also blended together; they gave the old man three hundred instead of five hundred rubles, and he forgot out of sheer joy to take the rein off. In many variants the buyer takes the rein by force. This is the case with Chubinskii (Vol. II, No. 103, 104), and Afanas'ev (Variant c). Often the buyer turns to those standing there for their approval and the public opinion determines that the rein should be relinquished together with the horse. Thus in the basic variant by Afanas'ev: "All horse dealers assail the old man: 'that is not customary, if you sell the horse, then sell the rein also." The old man is compelled to give in. In approximately this manner the sale is described in the variant which is printed in the *Zhivaia Starina* (1895, III, IV). In the tale by F. Aksamentov (taken from my transcription) the buyer (the magician) turns to the police and they force the old man to sell the horse with the rein.

Finally, in several variants we find the insult motif: in the transcription of Sadovnikov, in the collection by Afanas'ev (Variant c), and in the Viatka collection of Zelenin. This motif is carried out most completely in the last collection mentioned above: "The old man leads the boy through the forest. A raven crows from a birch tree. The old man asks the son: 'You have lived with Och, you must know what the raven is crowing about.' – The son fears that he will anger the father and denies it. The old man continues with his demand. Then the son says that the raven is prophesying a czar's kingdom for him: 'It is appointed that he shall become a czar and shall have his feet washed, and the father shall drink the wash water.' The father is not pleased and he threatens to sell the son once again. (He had not simply been apprenticed to Och, rather he had been sold.) Then the usual two sales follow. From the money thus acquired they live two years. At the beginning of the third year the old man again leads the son to market, remembers the old prophecy and decides to sell the son with the rein. 'He has angered me, thus I will sell him with the rein.' " (p. 125)[24]

This insult motif becomes in the tales of Vinokurova a driving force. But the action is not developed so directly as in the Zelenin variant or (even more so) in Sadovnikov's variant. The solution is gradually prepared. The entire episode, in spite of that which is externally marvelous, takes on a deep inner truth and convincing power.

In several variants the instances of the drunkenness of the old father appear. Thus with Chubinskii: the seller entertains the old man, and when the latter becomes drunk, he steals the rein. This simple allusion is developed by Vinokurova into a complicated scene, rich in psychological detail from real life. The old man already knows about the prophecy of the fortune-telling birds, but this prophecy had apparently made no impression on him. On the contrary, he calms the son who is ashamed to talk about it: "Well, it won't do any harm, it's not true anyway. Is it possible then that you will become a king" – he says to his son Mit'ka. But on the third day of the journey into town with his son (who has already changed himself into a horse) he sees a tavern along the way with its doors open. The old man, who had never been in a tavern, decides to go in. "What is so great about that; I have some money. I will go in and have a drink." He ties up his stallion and goes into the tavern. "Hey there, innkeeper, pour me a glass full!" The innkeeper does it, he drinks and it pleases him. "Pour me a second!" He stayed in the tavern for a long time. The drunk always has a lot to tell. The stallion begins to get impatient, *paws with his hoof in front of the tavern* – but the old man does not stop drinking until he is quite drunk. He comes out of the tavern, *whips the horse and tears at its bridle.* – "If I want to, I'll sell you with the rein for this rigmarole, you will wash your feet and I will drink your bath water!" – That's the way it is, a drunk is a drunk. He goes to the market and demands three hundred rubles for the horse, without the rein. The buyer asks: "Can't you sell it with the rein, grandfather?" – "Take it, for all I care!" and thus in his drunkenness he sold his son with the rein.

We see how the motivation in Vinokurova's rendition is deepened and made more complicated. In the variant from the Viatka collection the cause and the effect are both given together in the form of a simple reflex. That is the usual manner of folktale narration: a certain action and the reaction which follows immediately afterwards; all intermediacies, [and] diversions of action are missing. This is quite different with Vinokurova. She introduces into the tale a whole series of intermediate instances, among which she mentions some only fleetingly, but tells the next ones in detail, and as a result of their concatenation, builds up and develops her tale.

The feigned indifference of the father when he hears the prophecy, the gradual build-up of drunkenness, the anger about the dissatisfaction of the son, the derision of the drunk and the venting of his bad mood on the son, the drunken bragging and the threat, finally the surfacing of the sickness locked up on the inside which end with the conscious surrendering of the rein. Here we have the connected stages through which the narrator develops her tale. At the same time the action increases without a break, and the tale is continued in a forceful, tense tone – the drunken condition of the old man is presented in a series of monolog replies (not in the dialog). The dissatisfaction and unrest of the son is presented through a series of accounts of movements ("he paws with his hoof"), likewise the anger of the father ("he whips the horse and tears at the rein"). If in the variants of Zelenin and Sadovnikov this punishment of the son bears a purely external, folktale-like character, it takes on reality, deep truthfulness and convincing power in the case of Vinokurova. The purely external turning points with folktale-like characteristics become purely human experiences.

We must give mention to still one more detail. Usually the narrators forget the father right after the sale of the son. He appears anew on the scene only for the purpose of fulfilling the prophecy of the birds. Occasionally his growing poverty is mentioned. [25] Vinokurova is an exception to the common rule here too. Before she goes over to the presentation of the subsequent fate of the son who was sold, she remains for a while with [a description of] the moods and inner experiences of the old man. In rapid succession she tells us of his gradual sobering up, his grasping of what has happened, his remorse, finally the confusion and the futile searching for the son. "While still running through the city, his drunkenness fled him, he was afraid. – 'What have I done! Why did I sell him with the rein? Now I will never see my son again! Why did I go into the tavern and drink brandy?' He waited and waited for Mit'ka at the place where they had always met. But Mit'ka didn't come and didn't come. For an entire week he ran into town and always hoped to meet him somewhere. No, he doesn't meet him. And thus he had to live without Mit'ka."...

This episode deepens the dramatic situation of the tale and at the same time gives a touch of artistry by humanizing that which [in other folktales] is superficial fantasy.

The epilog of the tale is also very interesting from this

viewpoint. In the majority of Russian variants the end is with the traditional or even stereotypic marriage of the hero and the princess. In addition to this, however, the Russian folktale still mentions a female figure: the daughter of the magician, with whose help the pupil flees the vengeance of his teacher, the magician. This motif appears also in several West European variants – there it is connected with the subject of magical flight: "the magician's apprentice flees together with the girl." With Vinokurova the figure of the daughter of the magician is presented in more detail. She tells how the unfortunate apprentice has been hanging in chains for six months over the fire, he is completely smoked.

"Once the father (the magician) went away somewhere. His daughters said: 'Let's go in and see about Mit'ka.' The younger daughter entreats her older sisters to untie him and to give him something to drink." Out of fear of the father, the latter try to convince her otherwise, but she succeeds in carrying out her plan. "He walks, reels, stumbles, but the youngest sister feels sorry for him and she leads him to the river, gives him a drink, after she has freed him from the rein" . . .

Thus we see here two deliverers: tradition, however (which appears more clearly in the first part of the tale), requires marriage with one of them, and in fact with the second, the king's daughter. At the same time, however, the psychological and moral feeling of the narrator places special emphasis on the role of the first helper.

Vinokurova solves this problem in the following way: she has the hero return to the house of the magician and court his youngest daughter. Then he returns for some reason to the king. On the evening before the marriage the king's daughter, out of envy, poisons the bride, and thus the hero is compelled to marry the princess.

In this way the narrator averts the collision between the ending which appears natural and necessary to her and the ending forced upon her by tradition. It doesn't work, however: the entire episode appears to be forcefully shortened and has little motivation. But this artistic failure with which the narrator is affected, is characteristic and indicative. It underlines still more the basic tendency of her creation, of her artistic task.

This same artistic technique, capturing and presenting the internal reality of a happening, in order to throw light on its psychological detail, we can observe in the tale of The Unfaithful Sister. The usual scheme is the following: the sister, urged on by her lover (be it a robber, a dragon, a magician, or in the case of Vinokurova, a forest demon) sends her brother out to carry out difficult tasks, at the completion of which he would necessarily have to die. With the help of various magical animals delivered to him, the brother successfully overcomes all hindrances, kills the lover and devises a punishment for his sister. This kind of punishment is generally repeated quite uniformly in all variants.

If one pursues these episodes in Russian tradition, they appear in this sort of form: in the Krasnoiarsk collection[26] the brother chains his sister to a post and places a five bucket vat next to her: "If you fill the vat with tears, then I will believe you." It is the same with Rasdol'ski (No. 36) and with Zelenin (Viatka collection No. 6). This last tale, however, contains unusually gruesome details: the sister is hanged with her head down on a supporting girder. With Afanas'ev the details are a little different (No. 118, Variant c): Ivan the czar's son locks up his sister in a stone tower, places a bundle of hay over her and places two vats next to it, the one with water, the other one empty. "If you drink this water up, eat this hay and fill this vat with tears, then God has forgiven you and I forgive you too."[27]

This same type of punishment we also find in a more complicated form: two vats are put out which the unfaithful sister is supposed to fill with tears, the one for her brother, the other for her lover. Occasionally the punishment is formulated in a more complicated way through the choice motif: in Iavorskii's collection (where an unfaithful wife appears instead of an unfaithful sister) the husband puts out two vats as a test. The one is empty, the other is filled with coal. If she rues her deed then she will fill the first one with tears; if she is still worrying about the devil (her lover), then she will eat up all the coal. On the next day it turns out that she has wept no tears, but has eaten up all the coal, down to the last piece. Thus the husband commands his dogs to tear her to pieces and to throw her into the grave to the devil. This tale is also constructed in approximately this manner by (Chubinskii, Vol II, no. 48).

Afanas'ev introduces in his notes an interesting variant from the Bucovina: the hero digs three ditches. In each of the first two he places a barrel, in the third he buries his sister up to the waist. "You have," he says, "an evil heart, and you must repent. The barrel on your right shall be mine, and the one on your left – that of the dragon. I want to see which one you fill first with your tears." With that he wanders off into the distant world and returns only after a year. The barrel on the left proved to be filled with tears, the one on the right, however, remained empty. Then the brother buries his evil sister completely under the earth.

In the case of Vinokurova this tale follows, in its format, the variants just mentioned. It contrasts sharply with them, however, surpassing them in strength and splendor of presentation.

The brother led his sister to the place where he had slain her lover, the forest demon: "There is your lover!" *"She wept and wept, she rummaged and rooted in the ashes and found a boar's tusk, pressed it to her heart, wept over this tusk, this tusk of the forest demon."* Then the brother puts up two posts, between them he hangs a chest and puts his sister inside. Next to her he places two barrels. "There," he says, "weep a barrel full for me and one for the forest demon, then I'll let you go free. Whom will you first mourn, me or this tusk?" – *No, little brother! First I will weep for the tusk and next for you."*

In this manner the usual scheme of the subject is indeed presented in the tales of Vinokurova, but this scheme assumes a warm, living and human coloration with her – it takes on life and movement. Instead of the torpid figure of the unfaithful sister, the vivacious, powerful person of a woman filled with passion and grief appears before us. In this picture of confusion, presented in such detail, in which she presses the tusk of her lover to her heart, sprinkling it with tears, [and] in her proud and forthright confession one senses the living picture of a loving and suffering woman, not the marionette figure of epic [folk] poetry.

Here we certainly have no conscious calculation before us: no definite conscious goals guided the narrator as she was creating this picture, but here that unconscious, artistic drive asserts itself, that drive which under other circumstances, in other surroundings would have made her a great writer.

This art of creating psychological pictures one could illustrate with a whole series of examples. I will give one such [example]. In the tale The Czar's Daughter Freed by a Soldier, the hero has been left behind in the kingdom of the dragon by deception, and [he] is then carried by a bird into an unknown kingdom. There he hires out as the watchman of a house and a garden. The employer makes him a colonel: "You are a colonel! This will be your assignment: do not climb up into the highest old story of the old house, and don't climb down into the cellar, otherwise it will not go well with you: you will receive no reward and will have to serve three more years. And even if you go there unnoticed by anyone else, I will learn of it for sure."

The soldier-turned-colonel, however, could not stand it in the end. "Only one month in the year remains. Then it strikes him all of a sudden: 'What kind of soldier am I, what kind of a colonel, if I submit to the order. I'll have a look.' And he had a look." – "The boy who takes care of the house and the garden with him, runs toward him quite frightened: 'Why have you dared to go in there? The master will find out about it and you will have to serve three years for nothing.' – 'All right, that's no misfortune if I have to serve three years for nothing, but I did want to fulfill my wish. Let him find out.' Soon the master came back. – 'Well, fellows, how is it going, how is life here?' The colonel was not afraid. 'We will continue to live just as we have lived. Forgive me, I have violated your wish and I have been in the highest story of the old house and in the cellar. *You can have me hanged and beheaded, but the fulfilling of my wish is of more importance to me.'* And the master was pleased by this honest confession."

VI

As we have already mentioned, Russian researchers emphasize chiefly the importance of the biographical element. In research the problem of the agreement of the general tone of the tale with the individual character of its bearer is advanced into a primary position. With the application of these basic tenets, we must nevertheless observe certain limitations.

It is obvious during the recreation of the tale which takes place with each new telling, that personal experience and the narrator's own personality are of great importance, and that the coloration of the tale and the characteristic presentation of many local details are very dependent on these, [i.e., the narrator's personal experience and his own personality].

But can one decide with the required exactitude, *what* must be placed on the one hand under personal elements, and on the other hand under general poetics of the folktale and local tradition? Let's take an example. That the narrator [Vinokurova] worked in her early youth in the city as a servant girl, certainly had a major influence on her tales. Quite frequently a servant girl appears in them, most often as a chamber maid, and she certainly plays an important role. The maid appears to her to be necessary in every household. Through the maid the mistress learns of the new way to ask for alms ("not for the sake of Christ, but in the name of the eagle prince"); the *koshchei* has someone fetch the musically gifted servant girl; the chamber maid tells the father about the bad behavior of his daughter (The Tale of the Merchant's Daughter and the House Servant), and one servant girl performs a most important service for her mistress, a service for which she herself must later pay (The Tale of the Faithful Wife). Finally, a servant girl appears as the main character in the first part of the tale The Cursed Garden. Coachmen, cooks and house servants also play important rolls in the tales of Vinokurova.

In the case of those tales with servants [who play important roles] it is difficult to separate clearly the personal elements from the artistic. If in the Tale of the Eagle Prince the picture of ironing clothes comes completely from personal reminiscences, one can certainly look upon the presence and role of the servant girl in the Tale of the Faithful Wife as belonging completely to the original theme. One could explain the special emphasis on these roles in the renditions of our informant, as well as several realistic traits employed by her, by means of the personal element.

In the tales of Vinokurova the poverty motif assumes an important place, which she usually presents in tones of warmest sympathy and of deep compassion. In the tale The Promised Son, the old forester charges his foster son: "If someone, perchance a poor soul, fells a tree, don't [try to] squeeze money out of him, just forgive him his transgression without asking anything for it. I always do it that way with poor people. Father czar has forests aplenty. One must not offend the poor." – In the Tale of the Body-eating Woman, the merchant decrees at death: "And now, my dear wife, when I die, destroy the debt book of the poor, but record the debts of the rich." Then he commands further: "Three times yearly you shall arrange rich funeral feasts in memory of me. Now, after a half year and at the end of a full year, each time prepare a hot meal and invite everyone, so that all will be full: young and old, rich and poor. And give the poor as much as necessary."

The old father only succeeds in recognizing his son (The Magician and his Pupil) when the daughters of the magician reveal to him the secret. The latter explain their actions by means of their sorrow over the poverty of the old man. Thanks to their pity on the poor woman (in the Tale of the Merchant's Daughter and the Robbers), the heroine learns the secret of the robber's den, and she is able to save herself.

For the characteristic way in which Vinokurova usually treats a theme before her, the most revealing is the tale she told of the body-eating woman. The tale is a contamination of two stories from the *1001 Nights* which the informant in some way had the opportunity to hear, perhaps during her stay in the city: The Story of Sidi-Numan (first smuggled into *1001 Nights* by Golland) and The Story of the Porter and the Three Sisters.

The text of the Vinokurova tale is very close to the text of the two tales mentioned from the *1001 Nights:* she even retains such details as the arrival of the one-eyed people, the punishing of the bridegrooms who were turned into dogs, and so forth. Nevertheless, the tale told by Vinokurova can not be looked upon as a simple retelling. Her variant is, in the truest sense of the word, an artistic reworking in which Scheherazade's episodes and themes serve only as a base.

She sets forth the colorful fantasy of the oriental tale, true to her artistic method, she dwells with love and special care on the details of daily life, takes them up and explains them in her own way. She asks, for example, Scheherazade's porters for permission to take part in the feast in the following way: "You are alone, without men, who could entertain you with their talk . . . the women, however, do not understand how to spend their time and be happy without men," and so forth, thus the argumentation which the informant from the Lena places in the porter's mouth bears a completely different character: "Let me come and see your festival. I grew up in poverty and have never seen the likes of this."

In this motif (the motif of poverty) we again encounter the concurrence of instances of personal biography and personal sensitivity with cases of traditional poetics. This inseparable union appears especially clear in the Tale of the Faithful Wife, where the merchant's son marries the beggar girl who put him in chains. From this old, well-known subject Vinokurova creates an entire poem on poverty.

Collectors and researchers have often mentioned the special characteristics of feminine folk poetry. Onchukov and later the Sokolovs mentioned the loving portrayal of woman's life by their female narrators. In addition the Sokolovs point out the prevalence of feeling, of daintiness, of tenderness and of a sensitive tone in the tales told by women.

We can see this same thing in the Vinokurova tales. I have recorded 26 of her folktales, and among the multitudinous materials of her stories, female themes predominate noticeably: "the faithful wife," "the wise woman," "the merchant's daughter and the robbers," "the merchant's daughter and the household servant," "the wife of the minister," "the unfaithful sister," "the body-eating woman," and others. Seen from this viewpoint her repertoire is very valuable and complete. Even if the emphasis lies elsewhere, she puts the female theme in the foreground and treats it carefully, even if it is not the main concern. Thus, for example, in the tale of the eagle prince, much attention is given to the relationships between the *koshchei* and the abducted queen; in the Tale of the Prodigy – [there is much about] the history of his mother, likewise in the Tale of the Cursed Garden – [there is much about] the mother of the three warriors; in the tale The Magician and His Pupil the role of the daughter of the magician received special emphasis, and so forth.

The favorite theme of Vinokurova is the slandered, persecuted woman, the wonderful conception and birth of a son. The persecuted maid, fleeing from the unjust anger of the czar arrives at a beautiful island. On her way she comes to three trees with three wonderful apples on them. She eats these and feels that she has become pregnant. "In short, she gave birth to three sons." . . . One day they all come to their mother: Why were you in the forest, mother, and why did you give birth to us in the forest?" – She tells them everything in detail. "You see, my dear children, you were not born as the result of a mistake." – "Nor [is a] mistake responsible for the birth in the tale The Son Conceived by a Kiss.

The mother picture she always portrays with a special sympathy and delicacy. To be sure, we also find with her themes in which unfaithful mothers appear, for example in the tale The Milk of Wild Animals and The Goose with the Golden Eggs, but she handles them in her own specific way. The mother-punishment motif, which usually appears in the epilog of similar stories, we never find with her. In the Tale of the Goose with the Golden Eggs, the sons punish the lover of their mother, then leave her alone, leaving the punishment to the father. "There are no such laws, which allow for the punishment of the mother." And the father forgives her too.[28]

Usually, however, the mother portrayed with her is surrounded by a deep reverence, and she paints lovingly for us the cares of a mother: the education of the son and the mother's sorrow. She presents the mother picture in all its magnitude in the Tale of the Wise Woman. "Once there were three brothers, enormously rich." But the eldest separated from the younger brothers, became poor and died. The widow was left with her five-year-old son in poverty. And as poor as she is, she tries to make a good man out of this little son. When he is sixteen years old, the mother takes him to her in-laws. "Take my little Ivanushka, teach him thoroughly! I don't need money for him; my wish is simply that he learns to live with people." – Later the uncles slander their nephew to his mother. The latter returns home in deep sorrow. In the meanwhile the son arrives with his wife on richly laden ships, they disguise themselves in tattered clothes and go to the mother's house. The mother, however, sits there leaning against the window weeping. When the son identifies himself to her, she weeps still more. "My son, this is the way you came back to me! Wherever you have been [hanging around], go back there right away." The son asks urgently for a bed, but the mother, offended and disappointed in her hopes, angrily sends him to the outbuilding "where just turkeys and geese live." The son is hesitant about spending the night there. "Instead of spending the night in such a despicable place," he says to his wife, "we would rather return to our ship." The wife, however, does not agree to this. "That won't do, Ivan! Where your mother gave her blessing, there we must spend the first night." Just as moving and sympathetic is the episodic figure of the mother in the tale of The Faithful Wife, who is jealous of the honor and the dignity of her daughter.

These examples do not exhaust the supply of female figures and themes with Vinokurova. She dwells on various expressions of feminine feeling, and each time she approaches it with the psychological instinct peculiar to her. We mentioned above the mighty and passionate personality of the lover of the forest devil. Next to her the carefully drawn figure of the young beauty must be mentioned, who is whiling away her life at the side of her old husband. "Then Vasia comes wandering through the city, he goes walking (an exceptionally handsome young man he was). The minister's wife is sitting up on the third floor just then, she sees Vasia and sheds tears: 'Those are people! That is what one calls handsome! And I, at whose side must I waste away!' "

Among the favorite themes of Vinokurova, to which she often returns in her creations, are the prophetic dream and the music motifs. She believes in dreams, loves them and understands how to tell them. Once she told me one of her wonderfully poetic dreams, which reminded me instinctively of Turgenev's Lukeria. She confessed to me that such tales as the one with the untold dream, were her favorites. Most of all she loves the Tale of the Faithful Wife. We meet with the motif of the prophetic dream right away in the first three tales of the collection.

Her love for music and for sorrow came out most clearly in the tales of Vinokurova. The son of the eagle prince, disguised as a girl, casts a magic spell over the *koshchei* with his violin playing. In the tale of the body-eating woman "a fine, soft music" begins at the festival, then one sings "fine, sensitive songs." With great artistic energy and with great artistic enthusiasm she presents the playing of music by the magician (The Magician and His Pupil).

In the palace of the czar a party is being given. "There has been drinking and eating, and now music and dance are about to begin. Suddenly outside the window a simple peasant's balalaika[29] can be heard. A servant is sent to see what it is. He returns and reports: a man is playing on an unusual instrument. They listen, the music is pleasing to them. 'Quickly, call him into the room!' He is called. *Some laughed at his music, others wept, still others were pleased with it and started to dance. It seemed so beautiful to them, this music."* The poetic nature of the informant is expressed completely in this episode.

The appearance of the magician as a musician, of course, was not created as a motif by our informant. It is one of the traditional episodes of the folktale. One finds it also, for example, in the collection by Afanas'ev (Var. b). But the carrying out of this episode, the sensitive, artistic presentation of the playing itself, must, in my opinion, be attributed completely to the fine talent of Vinokurova.[30]

In conjunction with this the fine understanding of the informant for nature's beauties must be included. In her tales this could, of course, not be completely expressed. In traditional poetics the landscape plays only a secondary roll and is usually only superficially sketched out. Some researchers, for example E. Anichkov, even maintain, "descriptions of nature are completely lacking in folk poesie."[31]

The observations of Ėleonskaia seem to me to be deeper and closer to the truth. "The landscape," she writes, "plays only a limited role in the folktale; attention is paid to it only when the action being developed in the tale is influenced by it; thus a few careful strokes suffice to fix the external limitations of the event. The strokes are usually the same set and conventionalized [expressions] ("the steep mountain," "the thick forest," "the blue ocean" and so forth).[32]

Vinokurova is always trying to break through the limitations of traditional poetics. Her poetic feeling and her artistic talent draw her instinctively to nature descriptions and one can follow her attempts to expand the traditional framework, to destroy the obligatory norms and to find words and colors in order to portray her feelings for nature directly.

In a series of her tales we find only brief descriptions drawn with a few strokes: "They enter the garden, there is nothing unusual there. Withered trees, overgrown paths, nothing special . . ." (The Cursed Garden); Ivan, the czar's son guards the garden at night: "it is midnight, and it seems to him, as if his garden is getting brighter; it is getting brighter and brighter" (Ivan the Czar's Son and the Gray Wolf). But in other of her tales we find attempts at a more detailed description of the landscape picture.

Thus, for example, in the Tale of the Rich Marko, the oak is described in great artistic detail: "There stands an oak, moving back and forth, languishing, moving constantly back and forth." In the Tale of the Faithful Wife, a picture of nature in the spring is given: "The month of May arrived, the flowers in the garden began to bloom. He went walking with her. He walked and walked and sighed deeply. She asks him urgently: 'Why did you sigh so dissatisfied, what's wrong?' He: 'Oh, no reason.' But then he says: 'That's it, dear heart, when I was still a bachelor, I used to load ships about this time of the year and then sail on them out into the distant world. Thus I am sad now.' " – Here we see the suggestion of the unity of living nature with the inner life of the hero which is so characteristic for the Russian novel and the Russian novella. In connection with this direct feeling for nature one must also give several comparisons of Vinokurova, for example: "She whirls around him like a swallow, puts on his clothes and his boots, buttons him up . . ." (The Body-eating Woman).

I have already suggested that one can not always determine with the necessary clarity, in respect to individual details, what must be attributed to personal disposition and what to *the artist:* artistic instinct and feeling for style, – but the general coloration, the general tone of the style and of the personality almost always meet in the folktale. One can also observe this in the creations of Vinokurova.

The fundamental trait of Vinokurova's disposition is her sensitive, soft delicacy, and this soft coloration, this delicacy permeates all her tales.

Thus, in Vinokurova's tales for example, there are no obscene instances at all. She did indeed tell me the tale of the four Popes (more accurately: How the Peasant Cleverly Deceived the Entire Clergy), but this tale which I repeatedly recorded in the Verkholensk region, is scarcely recognizable in her rendition (in regard to manner, not in regard to content). There are of course several things in this variant which are not exactly proper for our urban ears, but she told me this tale; as it were, with a certain timidity, with an embarrassed laugh, as if she wanted to say: "That's the tale, that's the way it's told, it's not my fault." Whenever she has to touch on obscene themes or obscene happenings in her tales, she does it every time with an unusual tactfulness and, one can say, with a certain respectability. In the Tale of the Faithful Wife an officer comes to the heroine who has made a bet with her husband on her marital fidelity. "Now, and this guest comes to her. She takes him in properly and serves him. A guest is still a guest. After he ate and drank, he begins to get fresh, to say some shameless things, they were alone in the room. 'God protect me from doing such things as that!' She pulls away from him. He pulls a revolver from his pocket and threatens her, saying she would lose her life anyway. She is frightened and says: 'At the moment we are both dressed fully and that is so uncomfortable. When it gets dark, we'll go to bed together.' He decides to wait until night. 'But I will do that in the dark. Even when I married, I determined not to undertake these foolish things by the light of day.' "

Her humor is just as delicate and good natured as this. The daughters of the magician lead the tormented Mit'ka, who has been changed into a horse, to water. He is able to escape. "Just as he jumps into the water, with all four feet at one time, he took to his heels, and is still running! Only waves could be seen round about, as he swam away – and they jumped on the sister: 'Now, what will Papa say!' Fortunately Papa was just coming home." The humorous effect arises out of the usage of the endearing description "Papa" for a horrible, forest demon-like magician.

One last example. In one episode the shy, sensitive character of the narrator comes out quite clearly. In the Tale of the Faithful Wife which she loved especially, she describes the wedding feast in the following way: "All had drunk, the ladies and the caveliers are dancing and making fun of him: 'Why aren't you dancing with your young wife?' And he is afraid to ask her, perhaps she will feel sickened by this. How is she, the beggar woman, supposed to know how to dance? And he sends a servant girl to ask her." By the way, I would be hard put to determine whether we are dealing here exclusively with the delicate nature of the informant, or with the psychological instinct and tactfulness of the artist.

Thus the problem of the personality in tale telling definitely goes over into the problem of the artist.

NOTES*

1. No reference is given in the German text, but it appears that Azadovskii is referring to: A. F. Gil'ferding, *Onezhskie byliny, zapisannye A. F. Gil'ferdingom letom 1871 goda* [Onega Byliny, Recorded by A. F. Hilferding in the Summer of 1871], I-III. 4th. ed. Ed. by A. I. Nikiforov and G. S. Vinogradov. (Moscow & Leningrad, 1945-51).

2. A. D. Grigor'ev, Vol I (Moscow, 1904); Vol. III (St. Petersburg, 1910); A. V. Markov (Moscow, 1901); N. E. Onchukov (St. Petersburg, 1904). For further references see footnote number 6, page 9 of the Y. M. Sokolov book *Russian Folklore* (Hatboro, Pa., 1968).

3. The collection referred to here is the *Sbornik velikorusskikh skazok arkhiva russkogo geograficheskogo obshestva* [Collection of Great-Russian Tales from the Archive of the Russian Geographic Society] (Petrograd, 1917).

*4. Walter A. Berendsohn, *Grundformen volkstümlicher Erzählkunst in den Kinder- und Hausmärchen der Brüder Grimm* (Hamburg, 1922), p. 30.

5. Antti Aarne, *Leitfaden der vergleichenden Märchenforschung* (Helsinki, 1926), FFC No. 13.

6. No reference is given in the German text, but Azadovskii is apparently referring to Onchukov's *Sievernyie skazski* [Northern Tales] (St. Petersburg, 1908).

7. One of the heroes of the Kiev cycle, who derives mystical strength from Mother Earth. Known particularly as the slayer of the Nightingale Robber.

8. Reference here is apparently to Jahn's *Volksmärchen aus Pommern und Rügen* (1891).

9. The German word *Legende* suggests a story with some supernatural happening as its basic component. The German *Legende* is not to be confused with English legend (*Sage* in German).

*10. See O. Walzel, *Wechselseitige Erhellung der Künste* (Berlin, 1917) and *Die künstlerische Form des Dichtwerks* (Berlin, 1919).

*11. R. M. Volkov, *Skazka. Razyskaniia po siuzhetoslozheniiu narodnoi skazki* [The Tale. Researches on the Development of Subjects of the Popular Tale], Vol. I (Odessa, 1924).

*12. For more detail see V. M. Zhirmunskii, "Formprobleme in der russischen Literaturwissenschaft," *Zeitschrift für slavische Philologie*, Vol. I, 1-2 (Leipzig, 1925).

*13. Berendsohn, *Grundformen*, p. 129.

14. An artel is an association for common work.

15. (Helsinki, 1910), FFC No. 3. N. P. Andre'ev translated Aarne's catalog and enlarged it with the most recent collections of Russian folktales: *Ukazatel' skazochnyk siuzhetov po sisteme Aarne* [Index of Subjects of Tales According to the System of Aarne] (Leningrad, State Russian Geographic Society, 1929).

16. Koshchei the Deathless. "The meaning of this name is very hard to determine. There are at least three disparate ideas involved. First of all the most ancient is that which occurs in the Word of Igor's Armament, in which the word Koshchéy is used for a warrior of the hostile Pólovtsy; and, when Igor is said to be put on a Koshchéy saddle, it means he is taken into captivity. Hence the word *koshchéy* came to be used in Russian as meaning a slave, or a groom, originally a captive slave from the Pólovtsy who fought the Russians for over two hundred years. Consequently the word has a meaning in Russian folk-lore which has a widespread Aryan notion, that of a fearful Enchanter who lives in a mountain fastness far removed; runs away with the beautiful princess, and can only be slain by the valiant lover, going through unfordable streams, impenetrable forests and unpassable mountains, so as to catch hold of his soul which is contained in a casket, or in some other manner is always terribly enclosed. He takes this soul, which is as a rule lastly contained in an egg, up to the Monster's palace, scrunches it in his hand, and the monster dies. Thirdly, the word became confused with *kost'*, bone, and so came to mean a skeleton or miser, and a wandering Jew. The epithet 'deathless' does not mean indestructible, but that he can only be slain in an extraordinary manner and will not die in a natural way." Appendix note from A. N. Afanas'ev, *Russian Folk-tales,* trans. & ed. by Leonard A. Magnus (London & New York, 1915), p. 342.

17. "izbushka na kur'ikh nozhkakh"

*18. E. Rohde, *Der griechische Roman und seine Vorläufer* (Leipzig, 1876), pp. 414-415.

*19. E. N. Ėleonskaia, *Great-Russian Tales from the Province of Perm* [Collection by D. Zelenin] ("Influence of Locality upon the Tale"), *Ethnographic Review,* No. 1-2, 1915, p. 37.

*20. *Russische Volksmärchen,* trans. and introduced by August von Löwis of Menar (Jena, 1921), in the series *Die Märchen der Weltliteratur.*

*21. One must point out here Vinokurova's depiction of things extraneous. She is able to *see* the object which she is describing. A few examples: "A boy in a short coat and wearing a black cap came running up to him from somewhere . . ." "She emptied the glass, and then her color changed, she became more beautiful, even nicer . . . ."

*22. In Russian variants he does not change into a dog, but we do see in several variants a change into a bird.

*23. Johannes Bolte and Georg Polívka, *Anmerkungen zu den Kinder- und Hausmärchen der Brüder Grimm,* Vol. II (Leipzig, 1915), p. 61.

*24. In the tale by Sadovnikov the insult motif displaces the motifs of the return of the son and the new sale. The old man forces the son to tell him what the geese are talking about to each other. The son, after much hesitation, finally tells the old man: "They are saying this: when we come home with you and enter the new hut, mother will pour water on my hands and you will hold the towel for me." The old man didn't like what the son said, and he said: "Are you then my master and I your servant?" He crept quietly and pushed his son with God's help – kerplunk! – into the Volga." – "So, now I'll hold the towel for you! It's better than going begging!" (D. Sadovnikov, *Skazki i predaniia* *samarskogo kraia* [Folktales and Traditions from the Samara Region] (St. Petersburg, 1884), p. 214.

[It is] somewhat different with Afanas'ev (Var. C). After the father has gotten his son back, he goes home. A flock of wild geese flies over their heads honking. The father asks what they are making so much noise about. When the son answers he doesn't know, the father pushes him into the river, angered that the son has studied so long and has not learned anything. (Vol. III, p. 289).

*25. In several variants the father sets out in search of the son and helps free him – e.g., Aksamentov (my [i.e., Azadovskii's] transcription).

*26. *Zapiski krasnoiarskogo pod'otdela vostochno-sibirskogo otdela* [Records of the Krasnoiarsk Division of the Eastern-Siberian Department] Russian Geographic Society, Vol. II (1906), p. 122, No. 29.

*27. Other forms of punishment: he shoots his sister (*Permian Collection,* No. 41); he blinds her (*ibid.,* No. 5); he decapitates her (Afanas'ev, No. 118, Var. I); he binds her naked to a tree (*ibid.,* basic variant); she is pecked to pieces by birds (Afanas'ev, No. 118, Var. d); he leads her into a dark forest, binds her to a tree trunk with her head downwards, lights coals and places them under her head (Sadovnikov, No. 11).

*28. We find this strict punishment of the unfaithful wife quite often, however. Similar strict punishment befalls the unfaithful sister.

*29. A traditional mandolin-like stringed instrument.

*30. As a comparison I offer this episode from the Afanas'ev collection: the teacher . . . "takes some string music, crosses the street and plays. The czar's daughter Marfida says to the czar: 'Little father, can't we have him come in?' The czar sends for him. And he came and played on his lute and pleased the czar and his daughter, and came a second and a third time. The czar asked him: 'How shall I reward you?' and so forth!" (Afanas'ev . . . Moscow, 1914, Vol. III, p. 288).

*31. See F. Weber, *Märchen und Schwank* (1904), p. 32.

*32. E. N. Ėleonskaia, *Velikorusskiia skazski permskoi gubernii* [Great-Russian Tales from the Province of Perm] (Collection by D. Zelenin): "Influence of Locality upon the Tale," *Ethnographic Review,* No. 1-2, 1915, p. 39.

*The footnotes marked with an asterisk are in the original German text (FFC No. 68). Although the numbering has been changed in this translation, they are still in the same order and correspond to the original 20 footnotes of the German text.